T0140538

Topics in Intelligent Engineering and Informatics

Volume 10

Series editors

János Fodor, Budapest, Hungary
Imre J. Rudas, Budapest, Hungary

More information about this series at http://www.springer.com/series/10188

Ryszard Klempous · Jan Nikodem

Editors

Innovative Technologies
in Management
and Science

 Springer

Editors
Ryszard Klempous
Wrocław University of Technology
Wrocław
Poland

Jan Nikodem
Wrocław University of Technology
Wrocław
Poland

ISSN 2193-9411   ISSN 2193-942X   (electronic)
Topics in Intelligent Engineering and Informatics
ISBN 978-3-319-36359-2   ISBN 978-3-319-12652-4   (eBook)
DOI 10.1007/978-3-319-12652-4

Springer Cham Heidelberg New York Dordrecht London

Printed on acid-free paper

Springer International Publishing AG Switzerland is part of Springer Science+Business Media
(www.springer.com)

Foreword

This Springer book series is devoted to the publication of high-quality volumes that contribute to topical areas related to intelligent engineering and informatics. The publication focuses on Innovative Technologies in Management and Science. It consists of eleven chapters that integrate ideas and novel concepts in Intelligent Systems and Informatics. The chapters were selected from the initial contributions to the World of Innovation Conference held on April 3, 2012 in Wrocław, Poland. World of Innovation themes spanned nanotechnology and IT innovations, especially as related to cloud computing which is the major thrust if collaboration between the Wrocław University of Technology and IBM Corporation, IBM Haifa Research Lab. The editors asked the authors to submit seven extended versions of their conference papers. Additional papers were solicited from the University of Arizona. Warsaw University of Technology, University of Applied Sciences, Hagenberg, Austria, and the Polish-Japanese Institute of Information Technology, Bytom, Poland. All manuscripts were subject to a stringent, anonymous review process and underwent revisions before the inclusion in this volume. Since it is said that both diversity and openness are key to innovation, this book truly embodies it. Owing to the skills of the editors, you can feel the flavor of exciting fusion. I am confident you will find interesting insight in the chapters that follow.

Tokyo, December 2013   Director of Career Advanced Professional School,
Professor of Graduate School of Decision Science,
Tokyo Institute of Technology
Junichi Iijima

Preface

This book has grown around a selection of papers from the WORLD OF INNO-VATION, (April 3rd, 2012, Wrocław, Poland), the conference concerning the latest achievements in the IT field that was organized by Wrocław University of Technology (WUT) in cooperation with IBM Global Services Delivery Centre in Wrocław. There were more than twenty papers in the conference and selected authors were invited to submit the extended versions of their conference papers.

This book includes updates and extended versions of 7, carefully selected contributions to the WORLD OF INNOVATION event (4 from WUT and 3 papers from IBM institutes). The rest of chapters were submitted by invitation from: University of Arizona, Tucson, USA; Warsaw University of Technology; Poland; University of Applied Sciences, Hagenberg, Austria; Polish-Japanese Institute of Information Technology, Warsaw.

Innovative Technologies in Management and Science represent a widely spread interdisciplinary research area with many applications in various disciplines including engineering, medicine, technology, environment, among others. Present edited volume is a collection of 11 invited chapters written by respectable researchers and experts in the fields. Chapters are focused on research and development of the latest IT technologies, in the field of Cloud Computing, IT modelling, as well as optimization problems. We may easily assume that Cloud Computing will play an outstanding role in the near future as it is useful, flexible and effective. Cloud computing can provide organizations with the means and methods needed to ensure financial stability and high quality service. Of course, there must be global cooperation if the cloud computing process is to attain optimal security and general operational standards.

The chapters presented here can be grouped into three categories: Innovation supported by Clouds Technology, Innovation proposals in management area, Theoretical refinement for innovative solutions.

The first part, Innovation supported by Clouds Technology is devoted to the foundational aspects of Cloud computing. It consists of 3 chapters that include:

Chapter 1, *Innovation that matters - the IBM Research way* by Oded Cohn from IBM Research Center, Haifa, focuses on IBMs vision for a smarter planet as well as cloud computing as a collaborative innovation platform. And further illustrates how IBM drives "innovation that matters", to the world.

Chapter 2, *Clouds in Higher Education* by Mladen A. Vouk from North Carolina State University (NCSU), through the lens of Virtual Computing Laboratory (VCL), discusses the needs, possible needs, and future of cloud computing in higher education.

Chapter 3, *Evolution, not revolution* by Mariusz Kozioł from Wrocław University of Technology, applies the revolution and evolution rules to computer architecture, computer systems and information technology history. Firstly, is given a short description of IBMi foundations and history. Secondly, is a description of IBMi Series introduction into academic curriculum environment at Wrocław University of Technology.

The second group of chapters of this book contains contributions to Innovation proposals in management area and consists of four chapters (4-7).

Chapter 4, *A Smart Road Maintenance system for cities - an evolutionary approach* by Hari Madduri from IBM Research Austin, TX, exemplifies the Smarter Planet/Smarter Cities initiatives being promoted by IBM worldwide. In this chapter is given a description of the created an innovative solution for road maintenance problems challenges currently faced by many cities in road maintenance.

Chapter 5, *Cloud IT as a base for virtual internship* by Jerzy Kotowski from Wrocław University of Technology and Mariusz Ochla from IBM Center, Warsaw, presents the University Cloud Computing Centre, created in 2010 by Wrocław University of Technology and IBM Poland. Main objectives of this, first in Poland Centre are: usage of the cloud for education, usage of the cloud for a Remote Educational Internship program, promotion of research on cloud technologies and local and international collaboration with other faculties and business partners in cloud services.

Chapter 6, *Development of intelligent eHealth systems in the future Internet architecture* by Paweł Świątek, Krzysztof Brzostowski, Jarosław Drapała, Krzysztof Juszczyszyn and Adam Grzech from Wrocław University of Technology. The personalization of eHealth systems is one of the most important directions in the development. There is given a description of few service-based applications utilizing a Future Internet architecture which supports quality of service (QoS) guarantees for the communication between their components.

Chapter 7, *Understanding non-functional requirements for precollege engineering technologies* by Mario Riojas, Susan Lysecky and Jerzy W. Rozenblit from University of Arizona, Tucson. Taking into account tests and assessments results in local middle school, authors founded that the elicitation of non-functional requirements for precollege learning technologies can be better understood by dividing schools in clusters which share similar resources and constraints.

The third part of this book is devoted to theoretical refinement for innovative solutions. This part is a very important section of the volume because the reader could find in it a wide range of fields where innovative solutions need essential theoretical support. This part deals with: logic controller synthesis, metaheuristic algorithms, and TV-Anytime cloud computing concept (chapters 8 -11).

Chapter 8, *FSM-based logic controller synthesis in programmable devices with embedded memory blocks* by Grzegorz Borowik, Grzegorz Łabiak, and Arkadiusz Bukowiec from Warsaw University of Technology, devoted to methods of design and synthesis for logic controllers in novel reprogrammable structures with embedded memory blocks. Proposed architectures offer ability to update the functionality, partial reconfiguration, and low non-recurring engineering.

Chapter 9, *Virtualization from University point of view* by Tomasz Babczyński, Agata Brzozowska, Jerzy Greblicki, and Wojciech Penar from Wrocław University of Technology, presents some problems connected with cloud computing and related to virtualization. Authors present network simulator with nodes, done in virtual environment and scheduling algorithm for cloud computing.

Chapter 10, *Metaheuristic algorithms for the quadratic assignment problem: performance and comparison* by Andreas Beham, Michael Affenzeller, Erik Pitzer from University of Applied Sciences, Hagenberg, Austria. The main goal of this chapter is to compare the performance of well-known standard metaheuristics with specialized and adapted metaheuristics and analyse their behaviour for quadratic assignment problem.

Chapter 11, *TV-Anytime cloud computing concept in modern digital television* by Artur Bąk and Marek Kulbacki from Polish-Japanese Institute of Information Technology, Warsaw. The aim of the TVA standard is to enable a flexible use of TV across a wide-range of networks and connected devices. In this chapter authors provide an overview of global standardization approach, namely the TV-Anytime (TVA) standard. They describe the motivation of its use, its principles as well as the base features defined by this standard to satisfy the most recent TV domain trends.

We would like to express our special thanks to the reviewers of this book: Carmen Paz Suárez Araujo, Jerzy Rozenblit, Zenon Chaczko, Jakub Segen and Johann Heinzelreiter.

We would like to thank the Foreword author - Professor Junichi Iijima from Graduate School of Decision Science, Tokyo Institute of Technology for his insightful and meritorious work in evaluating the presented chapters and for the final product of this outstanding selection.

The conference was made possible through the efforts of the organizers from Wrocław University of Technology, led by vice-Rectors, Professor Cezary Madryas and Professor Andrzej Kasprzak, as well as from IBM Global Services Delivery Centre in Wrocław.

Special thanks must go to Professor Tadeusz Więckowski the Rector of Wrocław University of Technology for his essential and financial support

Wrocław,
January, 2014

Ryszard Klempous
Jan Nikodem

Contents

Part III: Theoretical Refinement for Innovative Solutions

**8   FSM-Based Logic Controller Synthesis in Programmable Devices
        with Embedded Memory Blocks 123**
        Grzegorz Borowik, Grzegorz Łabiak, Arkadiusz Bukowiec

9   Virtualization from University Point of View 153
        Tomasz Babczyński, Agata Brzozowska, Jerzy Greblicki,
        Wojciech Penar

Part I
Innovation Supported by Clouds Technology

Chapter 1
Innovation That Matters – The IBM Research Way

Oded Cohn

Abstract. Innovation, a principal IBM value, is becoming more interdisciplinary, more collaborative, more global, and more open. This paper discusses the essence of collaborative innovation, which is built on four pillars: people, methodology, structures, and platforms. In addition, it covers the motivation behind innovation, followed by a discussion on its changing nature. Some examples from IBM Research, including the Global Technology Outlook process and services innovation, further illustrate how IBM drives "innovation that matters to the world". This paper focuses on IBM's vision for a smarter planet as well as cloud computing as a collaborative innovation platform.

1.1 Introduction: Why Innovate?

When asked what are the key factors necessary for success of any company, or what is the most important drive for growth, the majority of managers would rate "innovation" at the top of their list. Other factors such as resource allocation or global workforce management, which are clearly essential in running a company, have a lower priority.

Innovation creates new products, solutions and services that meet market needs, to keep the vitality of an existing business, which is essential for companies' survival. It may also lead to new businesses for the enterprise, hence more growth opportunities. Executives believe that customers need and will continue to demand innovative products and services. This means companies must strive to continuously offer innovative products and solutions to attract and retain clients.

Oded Cohn
IBM Research - Haifa, Israel
e-mail: cohn@il.ibm.com

© Springer International Publishing Switzerland 2015   3
R. Klempous and J. Nikodem (eds.), *Innovative Technologies in Management and Science*,
Topics in Intelligent Engineering and Informatics 10, DOI: 10.1007/978-3-319-12652-4_1

1.2 A History of Innovation - IBM Research

2011 was IBM's centennial anniversary, celebrating 100 years of on-going technology breakthroughs. IBM determined that its actions will be driven by three main values, one of which is innovation that matters, for the company and for the world. Perhaps this is best manifested by the company's corporate Research arm which, for years, has driven inventions that propel the company's growth and leadership in a changing world.

For over 60 years, IBM Research has challenged the technological status quo by exploring the boundaries of science and technology, bringing to light discoveries that have had a lasting impact on the world. Today, IBM continues to expand the frontiers of healthcare, energy, and telecommunications, to name a few. In a very real sense, IBM Research has altered the modern history of technological progress.

Some highlights of these inventions throughout the years are:

- *1944 Mark I:* IBM's first large-scale calculating computer developed in cooperation with Harvard University. The Automatic Sequence Controlled Calculator, or Mark I, is the first machine to automatically execute long computations.
- *1948 SSEC:* Selective Sequence Electronic Calculator is the first computer that can modify a stored program.
- *1956 RAMAC 305:* The first magnetic hard disk for data storage revolutionizes computing.
- *1957 FORTRAN:* A group of scientists at the Watson Scientific Computing Laboratory design a computer language called FORTRAN (FORmula TRANslation). Based on algebra, plus grammar and syntax rules, this becomes the most widely used computer language for technical work.
- *1964 SYSTEM/360:* The most important product announcement in company history to date.
- *1966 One-transistor memory cell:* This cell stores a single bit of information as an electrical charge in an electronic circuit.
- *1967 Fractals:* IBM researcher Benoit Mandelbrot publishes a paper in Science magazine introducing fractals.
- *1970 Relational Databases:* Databases allow individuals and companies to manage and access large amounts of data faster and more efficiently.
- *1971 Speech recognition:* IBM's first operational application of speech recognition is a computer that can recognize about 5,000 words.
- *1973 The Winchester disk:* The 3340 disk storage unit, also known by its internal project name "Winchester," becomes the industry standard for the following decade.
- *1979 Thin film heads:* These serve as an alternative for using hand-wound wire structures as coils for inductive elements.
- *1980 RISC architecture:* The Reduced Instruction Set Computer (RISC) architecture greatly enhances computer speed by using simplified machine instructions for frequently used functions.

- *1986 & 1987 Nobel Prize winners:* In 1986, Binnig and Rohrer each received a Nobel prize in the field of physics for their design of a scanning tunneling microscope. In 1987, Muller and Bednorz each received a Nobel prize for their discovery of superconductivity in ceramic materials.
- *1994 Silicon germanium chips:* IBM adds germanium to silicon chips, forming the basis of low-cost, high-speed transistors.
- *1997 Deep Blue:* IBM's RS/6000 SP high-performance super-computer became the first machine to beat the reigning world chess champion in a traditional match.
- *1997 CMOS 7S (Complementary Metal Oxide Semiconductor):* This is the first technology to use copper, instead of aluminum, to create circuitry on silicon wafers.
- *1998 Silicon-on-Insulator (SOI):* This high-speed transistors building process is used to deliver high performance microchips for servers and mainframes.
- *1998 Microdrive:* IBM unveiled the world's smallest and lightest disk drive – a potential boon to the digital camera market and to other consumer electronic devices that are demanding increasingly larger capacities for data storage.
- *2002 Millipede:* IBM demonstrates data storage density of a trillion bits per square inch, which is 20 times higher than the densest magnetic storage available at the time.
- *2002 Molecule Cascade:* The new scanning tunneling microscope produces an image measuring 12 nanometer x 17 nanometer.
- *2004 Blue Gene/L:* This is the latest addition to IBM's Blue Gene family of supercomputers.
- *2005 5-stage Ring Oscillator:* IBM identifies a material that can outperform silicon in terms of device density, power consumption, and performance.
- *2008 IBM is the first company to achieve a coveted supercomputing milestone:* The petaflop computer can perform 10^{15} floating point operations per second.

These Eureka points in time should not distract attention from the on-going work and continuous innovation that has been and will continue to be made through the years. The following advancements illustrate just a small subset of these continued breakthroughs in the area of disk drives

In 1956, the RAMAC 305, the first magnetic hard disk for data storage was introduced and revolutionized computing. The Random Access Method of Accounting and Control (RAMAC) permits random access to any of the 5 million bytes of data, stored on both sides of a 50 two-foot-diameter disk.

The "Winchester" disk of 1973 featured a smaller, lighter read/write head and a ski-like design that enables the head to float closer to the disk surface on an air of film 18 millionths of an inch thick. In those years, this achievement doubled the information density of IBM's disks to nearly 1.7 million bits per square inch.

In 1998, IBM unveiled the Microdrive, the world's smallest and lightest disk drive. This was a boon to the digital camera market and to other consumer electronic devices that were demanding increasingly greater capacity for data storage. The drive can hold up to 340 megabytes of data, enough to hold about 340 200-page novels.

For over five decades, IBM Research has devoted resources and scientists to the field of disk drives, and continued to innovate throughout the years. Just imagine, if we had stayed with the disk capacity and weight of the earliest days, today's laptop would weigh approximately 250,000 tons (And the name laptop would certainly not have come to be.)

With the world constantly changing and advancing, IBM has had to and will continue to innovate products and services to offer better solutions for its customers and the world.

IBM's Research division has led in technological breakthroughs for over 60 years. Over the years, the role of IBM Research has evolved from focusing purely on developing scientific advances applicable to technologies (that IBM could bring to market) to research that focuses on "here and now" business problems. These additions to the division's mission have led to a more open-minded, flexible, and collaborative way of thinking and operating. Through the practical application of today's research, the division not only helps IBM lead, but also helps define the way people interact and will interact with technology for decades to come.

Over the years, the focus of IBM Research has evolved right alongside the evolvement of IBM as a company. This started with core competencies in hardware and software, and has advanced into emerging services disciplines. One excellent example is the creation of Services Science Management and Engineering (SSME). SSME is a new academic discipline being spearheaded by IBM Research that brings together ongoing work in computer science, operations research, industrial engineering, business strategy, management sciences, social and cognitive sciences, and legal sciences to develop the skills required in a services-led economy. For the latest breakthroughs and innovations from IBM Research see [1].

Research plays a central role in developing integrated solutions, working across the different business units to combine hardware, software and services.

IBM Research recruits the most talented candidates from around the world, with nine labs located in the USA, Switzerland, Israel, India, China, Japan and Brazil.

1.3   The Shift in Innovation

Proprietary innovation continues to maintain a central role in organizations that wish to control their intellectual property (IP) for the purpose of gaining business value from their inventions. The patent system is far from being obsolete and patent rights are an important source of income for many companies.

Still, new trends are evident, and innovation is becoming more multi-disciplinary, collaborative, global, and open (see [2]).

Multi-disciplinary - Innovation unites perspectives from various disciplines to accelerate technological advancement. A commonly known product, the smart cellular phone, combines the disciplines of materials, electronics, communications, software, and even social studies. Universities around the world observed that knowledge creation and innovation frequently occur at the interface of disciplines, so recognizing that the importance of interdisciplinary research and education are crucial to future competitiveness.

Collaborative - Using a community-driven approach to problem solving, people in industry, academia, and government are crossing organizational boundaries. Peter Gloor of the MIT Sloan School of Management stated that collaborative innovation is the modern application of "swarm creativity", which entails

- Creation of self-motivated teams
- Collective vision
- Enabled by technology
- Common goal
- Sharing of ideas, information and work

Global - Innovation draws resources that on one hand overcome geographic constraints, but on the other hand are driven by geographical requirements. Clearly the Internet serves as the glue that is uniting globally diversified people into a single collaborative team.

Open - IP practices are changing to accommodate open standards. Open standards as platforms will be discussed in a later section.

1.4   The Four Pillars of Innovation

The essence of collaborative innovation is built on four pillars: people, methodology, structures, and platforms.

1.4.1   Innovating People: T-Shaped Individuals

The T-shape stands for people who are *broad* (horizontal) and *deep* (vertical) [3]. T-shaped people are interested in Teamwork. They are excellent communicators, with real world experience, deep knowledge or specialization in at least one discipline and systems area, and good team work skills. These characteristics are all important when interacting with others who are deep in other disciplines and systems areas.

T-shaped professionals also make excellent entrepreneurs, able to innovate with others to create new technology, business, and social innovations. They are adaptive innovators, and well prepared for life-long learning in case they need to become knowledgeable in new areas.

Consequently, organizations like to hire people who encompass these broad and deep T-shaped skills and experiences. IBM, for example, has been collaborating with universities around the world to establish a new area of study that incorporates service science, management, engineering, and design (SSMED). This discipline will help prepare computer scientists, MBAs, industrial engineers, operations research, management of information systems, systems engineers, and students of many other discipline areas to better understand how to work on multidisciplinary teams and attack the grand challenge problems associated with improving service systems.

1.4.2   Methodology and Processes

An organization that is keen on collaborative innovation should establish a detailed methodology and processes to pursue it. Below are some of the ways that IBM, as a company, and IBM Research in particular strive to achieve collaborative innovation.

IBM employees are provided with global online collaboration tools, ranging from business to technical aspects.

The IBM Institute for Business Value (IBV) [4] conducts primary research, publishes white papers, presents research findings at industry conferences, and writes bylined articles for trade and business publications. In addition, IBM continues to research and collect information on a series of CxO studies that conduct extensive surveys and interviews with senior executives around the globe including: CEOs, Chief Financial Officers (CFOs), Chief Human Resource Officers (CHROs), and Chief Marketing Officers (CMOs). The results are published every two years. For example, IBV in conjunction with the Economist Intelligence Institute (EIU) published the 2010 Digital Economy Rankings, a report that evaluates and ranks 70 countries in terms of their ability to use information and communications technology to the benefit of the country, its businesses, and its citizens.

The IBM Global Innovation Outlook (GIO) conducts a worldwide dialog on innovation, business transformation, and societal progress. A series of meetings are conducted with participants from worldwide renowned universities and industries, thus collaborating across a global ecosystem of experts. Naturally, IBMers participate as a minority group. The resulting reports uncover new opportunities and insights that will shape business and society. Examples from past years include focus topics such as: water, Africa, cities and more.

On the more technical aspects, IBM Research carries out an annual Global Technology Outlook (GTO) process . The GTO aims to identify emerging technology trends significant to existing businesses and with a potential to create new ones. The trends must be ground breaking, impact customers, and help the company understand customer challenges. The findings of the GTO have a direct influence on IBM's technical strategy and are used to initiate the major Research efforts in the coming year.

IBM ThinkPlace is a popular online community and portal for researchers to post their ideas and thoughts. Other people can read, comment and enhance the original suggestions.

Occasionally IBM holds an Innovation Jam. Technically, this is similar to Think-Place, but is accompanied by an intensive campaign for specific predefined subjects. IBMers are called to participate, and sometimes business partners, customers and even family members are asked to share their thoughts and comments. A miniature version of the corporate wide jams is the "idea lab", dedicated to one customer or partner who raises one or more topics, and seeks the perspective and knowledge of IBMers.

1.4.3   Structures for Global, Collaborative Innovation

IBM Research operates across the globe and organizes its labs on one-dimension, with strategies on a separate dimension. This organization is known as a matrix structure, which is of course not unique to IBM. Strategies such as software, services, and device technologies, draw their resources from the globally distributed labs. In this way, it is quite common to have a project executed in more than one lab. The researchers are equipped with the web-based tools to jointly and efficiently carry out the work.

1.4.4   Platforms for Collaborative Innovation

1.4.4.1   Open Standards

Open standards are vital for building platforms that accommodate contributions of innovative from various sources; these enable the different sources to work in harmony. In other words, the platforms create value for communities by connecting building blocks of various origins. In contrast, closed standards lock-in value for restricted groups.

These standards are openly documented, freely available, and evolve through collaboration in international standards organizations.

Figure 1.1 illustrates well known core information technology standards, such as the Linux open operating system, web services, XML, and more. These standards serve as building blocks of an infrastructure for interoperability.

1.4.4.2   Cloud Computing

The emerging concept and implementations of cloud computing have the potential to boost innovation by providing a platform for collaboration. This new consumption and delivery model is inspired by consumer Internet services and provides:

- Ease of provisioning: on-demand self service
- Ease of access

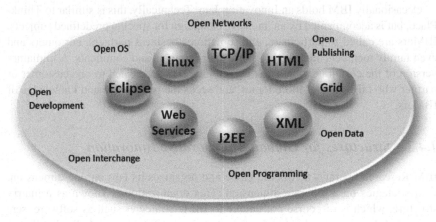

Fig. 1.1 Open Standards

- Rapid scaling and elasticity
- Reliability
- Sharing of IT resources
- Cost saving and low barrier to entry

Cloud computing supports innovation across several dimensions. It provides a platform for collaborating between partners, beyond the organizations defined fences. Other organizations, such as business partners, customers, and suppliers can work towards a common goal. Public affairs departments in governments and cities can use cloud computing to interact and collaborate with citizens, encouraging departments to innovate and improve the public services offered.

Using cloud computing, people can engage with each other in smarter ways, facilitating new and efficient business models. Global academic cooperation can run on the cloud infrastructure, so it advances research-intensive industries (e.g., life sciences).

The cloud provides an efficient and effective transition from research to deployment, which encourages the innovators by simplifying the route to 'real life' solutions. In-vivo experimentations are made possible. For example the Research Cloud (RC2) is actively used in IBM Research to experiment with new ideas from the IBM R&D community. Obviously, such a mechanism can be extended beyond the corporate boundaries (for nonproprietary projects).

The cloud itself is a subject of innovation, applied to all its layers. The layers are shown in Figure 1.2 below.

Here are some examples for the 3 layers:

- IaaS (infrastructure as a Service): In the managing of images of virtual machines, we apply advanced image re-mapping to maintain the continuity of operation.

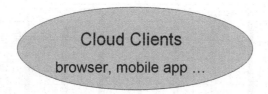

SaaS
Application: email, CRM …

PaaS
Platform: database, web server …

IaaS
Infrastructure: storage, servers, VMs …

Fig. 1.2 Cloud Computing - layers

- PaaS (Platform as a Service): We develop replication and consistency services, which are essential for managing stateful services and applications. Our approach is to come with a *consistent replication platform* to facilitate creation of highly-available elastic cloud services.
- SaaS (Software as a Service): We provide innovative de-identification (anonymization), which supports privacy by masking certain fields in documents, based on the viewer's authorization.

1.5   Innovation in Services

Service Economy takes a large percentage of countries' GDP (gross domestic product), as can be seen in the following chart:

As a result, innovation in services is vital to the world's economy.

Innovation in Service is needed for:

- Better modeling
- Data capture methods and tools
- Advanced analysis (machine Learning is very important here)
- Validation of business metrics (MBA skills are required)
- R&D for new and improved services assets. Asset based services are key to efficient provisioning, as opposed to just labor based services.

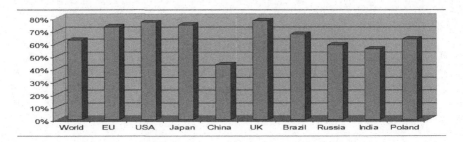

Fig. 1.3 Percentage of services contribution to the GDP

Outcome Based Services, whose purpose is to exploit the causal relationships between IT (information technology) capabilities and the business outcomes, is an area where innovation is also required. In other words, this is about expressing the improvement in a measurable business outcome (e.g., revenue, or profit) as a function of a specific investment in IT (e.g., more storage, improved software license).

Innovation is required in every step:

- Modeling of the business processes and the relation to business outcomes
- Data capture and analysis to predict results
- Evaluation of results according to business metrics
- Configuration and integration of assets to build a system that supports the business outcomes

In the following section we give an example of the application of outcome-based services to the area of customer churn, to predict which subscribers are likely to leave a telco. This example illustrates how *modeling, analysis*, and *metrics* are used.

Churn, or customer turnover, is a widely-recognized problem today for most mobile telecommunications providers. With customers being the telco's most valuable asset, gaining a more in-depth understanding of their behavior is a huge advantage, which in turn can help develop strategies to retain them.

The IBM solution starts by crafting operational models and value drivers to identify the key business capabilities that are important to the telco company under study. Typically the telco has programs like group retention, family bundle offers, loyalty offers, and more. Group retention will need social network analytics, loyalty offers need customer lifecycle management, and family bundles call for the customer's individual view. These are examples of the relations between the business operations and the IT tools, DB, platforms, and interfaces.

The team uses spreadsheets to analyze and create projections based on patterns learned through people's behaviors. Gaining an understanding of calling patterns, social behaviors, and how they evolve in such huge populations would normally require long-term research studies. With new smarter modeling and analytics, IBM can offer valuable knowledge to the telco operators. The results of the analysis are then used to decide on a solution approach. The solution is then configured with existing functional assets and deployed to support the Telecom's business outcomes.

In additional to customer churn, the analysis can be extended to support several other types of outcomes in the telecom industry. For example, these may include: increasing the number of annual activations, increasing the average revenue per user, boosting revenue from value added services, or reducing the ratio of call center employees/ subscribers (while keeping the same quality of service to the subscribers).

1.6   Smarter Planet: THE Multidisciplinary Innovation

Quoting from the IBM campaign on Smarter Planet: *"Every human being, company, organization, city, nation, natural system and man-made system is becoming instrumented, interconnected and intelligent. This is leading to new savings and efficiency-but perhaps as important, new possibilities for progress."*

IBM's vision for a smarter planet [5] calls for a multitude of innovations, crossing many disciplines.

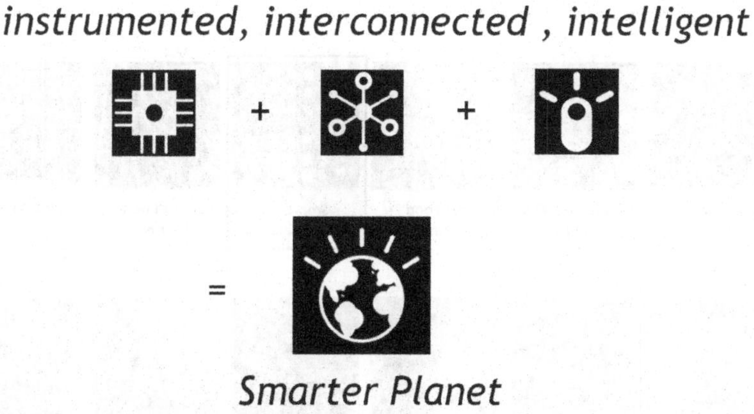

Fig. 1.4 IBM's vision of a smarter planet

Instrumented: The transistor, invented 60 years ago, is the basic building block of the digital age. Now, consider our world today, in which there are a billion transistors per human, each one costing one ten-millionth of a cent. This means we can instrument virtually anything, generating enormous amounts of data. Further, most of the people in our world are mobile phone subscribers, and the number of RFIDs is at least an order of magnitude higher. In short, humans and basically anything can be tagged, and act a sensor that emits enormous amounts of data.

Interconnected: Data is being captured today as never before, and we have the means to interconnect it. Data networks connect people, and in parallel with the Internet, there is an emerging Internet of things (where things, like people, are communicating with each other).

Intelligent: We want intelligence to be infused into the systems and processes that make the world work into things no one would recognize, such as: computers cars, appliances, roadways, power grids, clothes, even natural systems, agriculture and waterways. Once the interconnected instruments send the collected data, large and systemic patterns can be revealed of global markets, workflows, national infrastructures, natural systems and more, drawing conclusions and making decisions.

Since 2008, IBM has talked about what it takes to build a smarter planet. We have learned that our companies, our cities and our world are complex systems actually systems of systems. Advancing these systems to be more instrumented, intelligent, and interconnected requires a profound shift in management and governance toward far more collaborative and innovative approaches.

Smarter planet solutions are needed virtually everywhere; thus innovation is needed virtually everywhere. Figure 1.5 highlights some of the areas where smarter systems can help the planet.

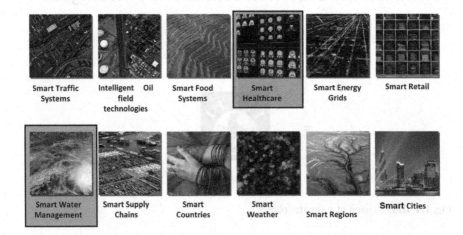

Smart Traffic Systems   Intelligent Oil field technologies   Smart Food Systems   Smart Healthcare   Smart Energy Grids   Smart Retail

Smart Water Management   Smart Supply Chains   Smart Countries   Smart Weather   Smart Regions   Smart Cities

Fig. 1.5 The Concept of a Smarter Planet - Solutions for Society

Two main examples, for the highlighted areas, are smart healthcare and smart water management.

1.6.1   Smarter Healthcare

IBM is engaged in many projects dealing with smarter healthcare, ranging from the electronic health records to scientific work done in medicine. EuResist project is a consortium of European universities, hospitals and industries, aimed at the development of an integrated system for clinical management of AIDS "drug cocktails." A drug cocktail is a mixture of medicines that should be adapted to the patient, based on the individual and his/her medical history and condition. Information from more than 33,000 patients was analyzed, and used to build the prediction model. By the end of the project, the world's largest database centered on HIV resistance and clinical response information was created. The system is accessible through web interface, and is freely available to the global medical community. The most important thing is the outcome: EuResist prediction engines are 76% more accurate than humans. The innovative work and the impressive medical results were achieved according to the 3 elements of the smarter planet paradigm: sensing a lot of data, connecting it to central processing, and infusing intelligence to make the most of it.

1.6.2   Smarter Water Management

Water is becoming a scarce resource in our world, and great efforts are being invested in generating more drinkable water. But once we have the water closer to its consumers, we lose a substantial part of it. It is estimated that 20% loss in European countries, with London frequently mentioned as a notorious example of 50% loss due to leakage in aging pipelines.

With the proliferation of electronic water meters (sensors) that transmit the measured flow to a central location where it can be analyzed, we can considerably mitigate this severe problem. Below are examples of IBM projects that apply to several management and maintenance tasks of the water system.

IBM has developed systems that have the ability to detect leakage at the water network level using analytics. The anomalies between modeled performance of the network and actual flow meter and pressure readings, are used to identify potential leaks. This is done non-invasively.

Water loss is reduced using dynamic pressure control. The water company can dynamically adjust pressure in real time so that only the minimal required flow is supplied. This in turn promptly reduces the water wastage (the lesser the pressure, fewer water goes through the hole), but also saves energy (in pumping) and reduces maintenance costs (the pipes live longer under the lower pressure).

When building a new water network, or repairing and augmenting an existing one, the IBM solution provides an optimal valve placement for pressure reduction. The network model is used to find the optimal location of valves that enables the most effective pressure management action. Referring to the previous projects, then for leaks in the pipes this can save water and avoid worsening the leak, perhaps into a full blow-out.

Asset management enables rapid repairs and maintenance planning and creates a seamless process of "detect a leak and fix it" reducing the cost and time needed to fix the problem. This system also enables optimized predictive maintenance to reduce capital and asset lifecycle costs.

1.7 Summary

After discussing why innovation matters, the trends in innovation and most importantly the collaborative nature of innovation, it seems only proper to end with the following words: "let's collaborate to innovate".

References

1. IBM Research; Explore our latest breakthroughs and innovations,
   http://www.research.ibm.com/featured/whats-new.shtml
2. Chesbrough, H.W.: Open Innovation: The new imperative for creating and profiting from technology. Harvard Business School Press (2003) ISBN: 1578518371
3. Word Spy; T-shaped,
   http://www.wordspy.com/words/T-shaped.asp
4. IBM; IBM Institute for Business Value,
   http://www-935.ibm.com/services/us/gbs/thoughtleadership/
5. IBM; Let's build a smarter planet,
   http://www.ibm.com/smarterplanet/us/en/?ca=v_smarterplanet

Chapter 2
Clouds in Higher Education

Mladen A. Vouk

Abstract. North Carolina State University (NCSU) has been operating a production-level cloud, called Virtual Computing Laboratory (VCL) since 2004. VCL is an award-winning open source cloud technology that at NC State delivers services ranging from single desktops (VDI), to servers to, high-performance computing (HPC) and data (HPD). It serves over 40,000 users and delivers about 250,000 service reservations per year along with 14+ million HPC/HPD CPU hours per year. Through the lens of VCL, this paper discusses the needs, possible needs, and future of cloud computing in higher education – the range and complexity of the services, the needed capabilities of the cloud architecture and implementation, and possible future development directions embodied in the vCentennial campus model.

2.1 Introduction

We know how to construct clouds [2, 3, 6, 8, 11, 12, 13, 14, 19]. It is less clear that we know how to use them, particularly in research-intensive higher-education environments where teaching, research and innovation need to co-exist in a proactive and collaborative way, and need to be ahead of the curve.

Cloud computing is now sufficiently well understood that real [15, 19] and de-facto [14] standards are beginning to emerge. Today a general cloud architecture is service-based and has at least three layers: Infrastructure-as-a-Service (IaaS), Platform-as-a-Service (PaaS), and Software-as-a-Service (SaaS). IaaS is concerned with (seamless and elastic) provisioning and configuration of the basic real or virtual resources, such as computational hardware, networking, and storage infrastructure. PaaS is concerned with a higher level set of building blocks such as specific operating system platforms (e.g., Linux or Windows), middleware provisioning, and

Mladen A. Vouk
Department of Computer Science, North Carolina State University, Raleigh, NC 27695, USA
e-mail: vouk@ncsu.edu

© Springer International Publishing Switzerland 2015   17
R. Klempous and J. Nikodem (eds.), *Innovative Technologies in Management and Science*,
Topics in Intelligent Engineering and Informatics 10, DOI: 10.1007/978-3-319-12652-4_2

general mid-level cloud services such as accounting, authorization and authentication, provenance data collection, etc. SaaS is usually limited to the use and development of specific applications, and higher-level services. In general this model allows for X-as-a-Service, where X can be anything that is relevant to the user or customer community, e.g., Security-as-a-Service, Cloud-as-a-Service, etc. How much access a user has to a particular cloud layer and how much control (e.g., can a user be root on the platform provisioned for the user, can user define VLANs, can user modify the application or only input data and get results from it) depends on the nature of the cloud, institutional policies, user privileges, and so on.

In [4] authors observe that *"within an educational and/or research environment, cloud computing systems must be flexible, adaptable and serve a wide spectrum of users from among students, faculty, researchers, and staff. For students, educators, and researchers to use cloud computing as an integrated tool for their work, it must be designed to deliver services that support everyone from a novice user up to the most sophisticated expert researcher. As part of the mission of a higher-education institution cloud computing needs to reliably deliver:*

- *A wide range of teaching and other on- and off-campus academic IT support services to students, teachers, and staff.*
- *Research-level computational, storage and networking services in support of the research mission of the university.*
- *Any other information technology services that the institution needs. This may include outreach related IT services, continuing education IT services, and IT services needed to administer the institution, such as ERP services."*

This is not surprising. Cloud concepts were not invented overnight. They evolved over decades from research in virtualization [1], distributed computing, networking, web, utility computing, grids, software services, autonomic computing, and so on [12]. Universities were at the center of many of these developments which were in many situations driven by the actual needs of the education and research of the higher-education institutions.

2.2 Capabilities

There is a basic set of capabilities that a good modern information technology system needs to support. Most of these capabilities are essential for adequate support of a full cloud computing environment that is capable of serving the broad range of needs a top research intensive university may have. This is illustrated in Fig. 2.1. Each bubble represents an essential capability, tool or function needed to support cloud layers from Hardware-as-a-Service all the way to Cloud-as-a-Service and Facilities-as-a-Service (or virtualized brick and mortar) [7].

Obviously computational, networking, storage and other physical resources are needed. Their utilization and management may be enhanced through virtualization and today most of the clouds do that. However, it is important to remember that

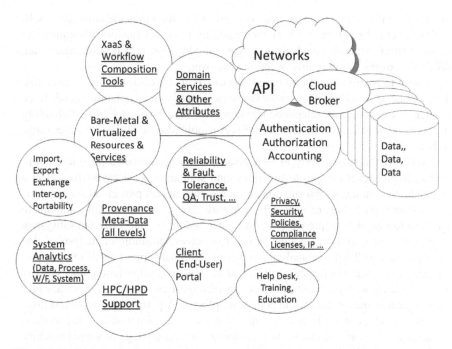

Fig. 2.1 Desired cloud subsytems, tools and capabilities

behind the scene there always are physical resources that need to be managed and
that anchor any set of information technology (IT) services. At a more complex
level, a cloud environment offers higher level service building blocks - composi-
tions (or stacks) of software consisting of operating systems, middleware, and applications running on those resources. Those building blocks can be composed into
more complex services and workflows. This is often a tedious, error prone and overhead inducing process. Therefore to construct cloud-based higher level services and
workflows, especially in the self-service mode, it is necessary to augment cloud environment capabilities with PaaS and SaaS composition and workflow component-
based construction tools [10, 9].

Naturally, a good environment needs to have an appropriate level of security,
privacy, and policy compliance capabilities, as well as capabilities to honor licensing and intellectual property boundaries. It also needs to provide a human user access through some type of interface (e.g. web, client-side graphical user interface or
GUI), but it also needs to be able to communicate (via application programming interfaces, API) with computer applications, with other clouds, and similar. The usual
authentication, authorization, accounting (e.g. for chargeback), and access control
services need to be there, as do customer-oriented help-desk, training services, and
trouble-shooting services.

As the economic model of cloud-based services and resources (at all
levels) evolves, it is reasonable to assume that some of those services may be more

affordable if they came from a commodity provider rather than the university itself. A flexible cloud solution will have the capability to broker functionally equivalent services from external providers to mitigate situations such as peak demand load, disaster recovery, or continuity of service fail-over.

Central to trustworthy cloud-based services is their reliability and availability. This can be achieved in many ways. Our experience is that availability needs to be similar to what classical telephone services offer (five nines, 0.99999). Reliability of the cloud with respect to security failures should also be in the five nines range or better.

To achieve that pro-actively, as well as satisfactory performance, provenance support of higher-level composite functions, and similar, it is necessary to collect a considerable amount of information about the current and past cloud status, state, performance, processes, data, system, software, users, and so on. This amounts to having a robust and comprehensive provenance and meta-data collection subsystem as part of the cloud infrastructure. In order to achieve meaningful analysis of that information a well balance analytics and computational support needs to be implemented. This then, for instance, allows us to perhaps predict failure and security states of the cloud components, pro-actively manage them, and provide privacy protection, license management, and policy compliance. A particularly important function of a cloud, even a private cloud, that may have users from all over the world is the ability to import and export data, computational codes (e.g., as virtual machine appliances or OVF files), and at the same time understand local, state, country and regional policies, regulations, laws and similar in order to effect appropriate import/export controls, manage cost optimization in the case of service brokering and the like.

In research universities it is also very cost-effective to have the cloud offer both services which are traditional (such as desktops, servers, clusters, etc.), and high-performance computing (HPC) and data (HPD) services (e.g., for simulations, big data analytics and similar).

There are, of course, many other fine-grained functionalities and specialized services that particular clouds may need to have. One may be very high grade security if the cloud is being used to host high-assurance applications and data, national security information, and the like. However, without at least the bubbles shown in Fig. 2.1, it may not be wise to get into the business of constructing and hosting a cloud.

2.3   VCL

2.3.1   Overview

Virtual Computing Laboratory (VCL, http://vcl.ncsu.edu) is an open-source (Apache, http://vcl.apache.org) cloud technology that was originally developed at North

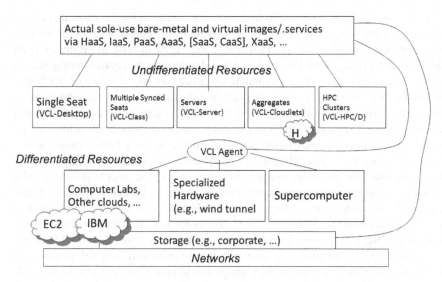

Fig. 2.2 NCSU VCL User services. Undifferentiated resources are malleable and can be used at IaaS or PaaS level, differentiated services may have a number of restrictions on how much a user can control the resource(s). They are typically SaaS, AaaS or higher, and may be attached to the cloud directly, or may be brokered by VCL from a non-VCL provider.

Carolina State University (NCSU), and has been in production use at NCSU since 2004. At NCSU VCL serves over 40,000 accounts and delivers about 250,000 service reservations per year, along with another 14+ million HPC/HPD CPU hours. The range of services offered by VCL is illustrated in Figure 2. At NCSU VCL is used to support individual students (typically augmenting their personal devices with software and services they may not have installed, e.g. specialized homework software), classes and labs, individual longer term servers and server clusters, storage, and of course HPC/HPD services. VCL capabilities cover almost all the bubbles shown in Figure 1, although for some of the services the user would need to have administrative privileges. Advanced users are allowed to create VCL images (software stacks based on NCSU licensed platforms such as Windows 7 or Linux) and deploy them to the user groups they are authorized to serve. VCL has a sophisticated role-based privilege management tree that can isolate individuals, groups and institutions, help ensure that licenses are honored, and in combination with other VCL and institutional services, can be used to implement a number of specialized policy, privacy, security, and topology related mappings.

For example, a user who wishes to construct a small virtual lab or lablet that consists of a web server, a data base server, and one or more computational nodes with specialized software (e.g. Hadoop-based), could do that in two ways. One would be to use VCL environment functionality associate with each image, and the other would be to use the VCL API. A combination of the two is also possible.

In the first case, the user would first make a reservation for the base-image (e.g. just Linux with no applications on it). This would be used to construct the anchoring or parent or control image. That image may need to have more than just an application on it (e.g. web server). It may need to have some control intelligence to know which other nodes (images running on either bare-metal or virtual machines) are in the environment. Then the user would construct a separate data base image (basically make a reservation of a base-line Linux image, and then install on it a data-base), and a separate computational image (with whatever application is of interest). At this point the user would have constructed and saved three images web + controller, data-base, and application. Now the user would (through VCL web interface) configure the controller image to be a parent, and the other two images to be children of this particular parent (one could add as many children as wanted e.g. five computational node images). From now on, when the user makes a reservation of the parent lablet image, all three or more images are loaded. Depending on how the mapping of the images onto hardware (real or virtual) is done, this small lab cluster could be tightly coupled (low latency network) or loosely coupled. In any case, when the reservation is granted, each image would find in its $/etc/cluser\_info$ file primary IP number of all its siblings and of its parent, that is the parent and the children know about each other. Following illustrates the content of the file

```
[vouk@bn19-36 etc]$ more cluster_info
child= 152.46.19.36
parent= 152.46.19.5
child= 152.46.20.78
child= 152.46.20.86
[vouk@bn19-36 etc]$
```

This enables the controller to find all its children, and the children to find their parent without fixed IP addresses (latter is of course possible but in a cloud it could be a limiting factor). Now a student who makes a lablet reservation gets, on demand, an independent personalized set of resources. Typical VCL reservations are about couple of hours (e.g. homeworks, class), but longer term and open ended resource reservations are possible and are being used for different types of projects.

If a dynamic cluster is needed, for example we would like the lablet above to dynamically extend its computational resources, the controller could use VCL API at run-time (after being deployed) to make (elastically) additional reservation of the computational node and deploy a larger computational sub-cluster that would serve the lablet.

In order to support applications as the one above, as well as services that range from individual desktops (or seats), to classroom or lab clusters, to server farms, to specialized sub-clouds and HPC or HPD services, VCL has a hierarchical set of built in functions that allow sophisticated manipulation of VCL resources, their scheduling, their monitoring, and their management. VCL architecture supports a distributed cloud and therefore its resources can be located in one data center, or over a number of data centers in different states and countries. Out-of-the-box VCLs very strong and sophisticated Hardware-as-a-Service (HaaS) and Infrastructure-as-a-Service (IaaS) capabilities are augmented with basic Platform-as-a-Service (PaaS) in the form of base-line images (operating systems, e.g. Linux, Windows). At NC State we also then provide centrally a certain number of Software-as-a-Service (SaaS) images (e.g. MatLab images or Maple software images). However, for most part it is up to users to extend the basic PaaS and SaaS offerings with their own services. When permitted, they reserve a basic PaaS image and install on it (provided they have appropriate licenses) software they wish, save the image(s), and then make that service available to their user community. Centrally, NC State maintains perhaps 30 or so PaaS and SaaS services, the rest of the 2,000 or so available service images have been constructed by authorized advanced users (e.g. professors, research assistants, students).

Of course, not all of the images are available to all users. Role based access control and image visibility is maintained using a variety of data sources for example class roles, research group membership, etc. This allows very fine grained mapping of individual services (individual images, or image groups) onto specific resource groups, specific user groups, specific schedule groups, specific management nodes and data centers, and specific privilege tree branches and leaves. Through this one can control many things computing power, licensing, service isolation, security, and so on.

Security typically involves use of VPNs to initially access VCL private cloud authentication and authorization, and later to access activated resources. IP address locking is used to make the resources visible to authorized user only, and watchdog time-out is used to manage inactivity, resource hogging, and possible denial of service situations. Also, traffic monitoring is used to identify potentially compromised edge-resources and trigger resource isolation if needed. VLAN-ing can be used to isolate resources provided appropriate VLAN-ing hardware equipment, or similar virtual solution, is available. Real-time and asynchronous dashboards and other tools are available to analyze and mitigate VCL internal performance and issues. Interested reader should consult some of the following articles and web-sites to obtain more information about VCL [13, 11, 6, 7, 20, 21].

2.3.2 VCL as an Operating System

At this point, almost 10 years after we have started working on a cloud solution which has arguably become the first academic production level cloud in the world, we believe we know how to build a secure and efficient cloud that is functionally suitable for the broad range of needs that arise in research universities. A natural question is do we stop there? What is next?

We have decided that a natural extension to cloud services of the classical type is the ability to plug into that cloud almost any device that may be useful in the university from a microscope (particularly an expensive one), to a genetic sequencer, to a networking laboratory, to a classroom. The intent is to either increase utilization of the hardware by making it available 24/7, or virtualize it and thus increase utilization through multi-tenancy, or emulate, or simulate normal part or all of the physical device, room, and even building. This would, in theory, enable us to allow authorized external users (students, faculty, researchers) to access these resources in some form on a 24/7 basis. This helps extend access options and turn resources and services that are often idle at night, or out of reach to distance education students, or unavailable to remote researchers and industrial partners for some other reason, into easy to access utilities. It would be like walking onto the campus with appropriate permissions of course and making use of sometimes unique university research and teaching resources. If done right, this could level the education field, and provide unprecedented advances to higher-education resources and sophisticated research partnerships.

This of course, is not a new idea. Remote controlled equipment has been around for a long time. NC State has been experimenting with this for years [5, 4, 16]. For example, Fig. 2.3 illustrates the 1999 instrumented backhoe that was installed in the Department of Civil Engineering at NCSU and was network-enabled in collaboration with Computer Science for remote manipulation. In addition to joy-stick operation of the equipment, remote access also involved stereoscopic cameras and sound so that operators would have good visual and audio awareness of the remote environment. It was demonstrated at an Internet2 meeting in Atlanta. At about the same time much larger facilities were also being turned into virtual labs over internet at other places for example the University of Hawaii and the Association of Universities for Research in Astronomy connected eleven leading observatories to Internet2 networks via the Mauna Kea Observatories Communication Network [17].

Today, almost any significant equipment does come with processors, IP numbers, and ability to access and control it from a distance. However, most of the large equipment is part of stand-alone solutions, and not necessarily part of an AaaS cloud offering, and this is understandable. Open questions are many. For instance, how safe and secure this is, how to schedule this, what are the pedagogical and other obstacles, what are user interface issues, how to ensure that such as service is very utility-like in both ease of use and reliability and availability (probably a few other trust related parameters), and so on.

To learn more we have decided to pilot several more complex options to see how they can be smoothly (ideally plug-and-play) integrated into an environment such as VCL as Application-as-a-Service or Instrument-as-a-Service. In this context, VCL serves as a campus-area PaaS and SaaS container environment essentially an operating system for services that are composed of physical resources, information technology resources, and remote and local users and provides scheduling, access, accounting control, monitoring and similar for higher level services that need those functions.

2.3.3   Virtual Campus

As part of its business continuity plans NCSU can launch 100+ virtual classrooms on short notice from its cloud computing solution [21]. This option currently uses

Fig. 2.3 Turn of the century NCSU experimental virtual laboratory featuring remotely manipulated teaching and research backhoe (Civil Engineering, Computer Science)

Blackboard Collaborate (formerly Elluminate) to provide classroom spaces in the event of a physical campus shutdown. We also have Internet Reactor Laboratories that are available to external academic institutions who wish to utilize the NCSU PULSTAR reactor to demonstrate nuclear reactor operations and kinetics for their students. This capability enriches academic programs at universities without research reactors of their own, and may be used to expand the educational opportunities for nuclear engineering students throughout the United States and internationally. [16]. A considerable number of our networking degree and cloud computing training is done using VCL hosted virtual lablets, and so on. However, we also have a number of facilities which could benefit from remote access option, but for variety of reasons do not have that capability yet.

This has prompted us to initiate a project we call vCentennial. It is an ambitious vision of change in education and research paradigm that would offer NCSU the ability to replicate its Centennial Campus services and functionality of this physical environment and its virtual avatars anywhere, anytime in the world using a cloud of clouds platform. The plan is to appropriately virtualize its award-winning Centennial Campus [18]. Centennial Campus is a small city made up of NCSU research, teaching and outreach facilities, entrepreneurs, academic entities, private firms, and government agencies. The path that we are taking is to systematically build on the state-of-the-art infrastructure and facilities of the Centennial Campus. This is illustrated in Fig. 2.4.

Centennial Campus facilities and buildings are new, but not all of them are yet smart. The process of adding sensors and an intelligent campus operations capability is in progress. Centennial networking infrastructure is first class fiber in the building risers and in the ground, 1 Gbps to the desktop capability, 4 to 10 Gbps among the buildings, latest wireless technology in all buildings and classrooms, and an experimental digital wireless canopy called CentMesh. What we are in the process of doing is extending VCL capabilities so that it can act as cloud operating system for Centennial facilities and equipment that we wish to plug in. This involves

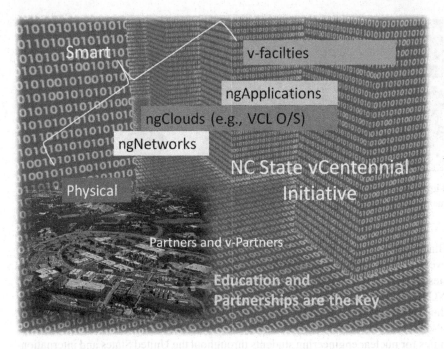

Fig. 2.4 The path to vCentennial digital campus

development of complex equipment interfaces, which may consist of one or more VCL images, in addition to the computational equipment attached to devices such as advanced electron microscopes or biomanufacturing laboratory and pilot production facilities, and interfaces to other campus laboratories and classrooms. The current challenge is that most of the complex applications (including some otherwise very sophisticated analytics software) are not fully cloud-ready. Next generation cloud applications will be aware of the applicable cloud services such as scheduling, authentication and authorization, provenance collection, policy management, and similar subsystems and would leverage such services (see Fig. 2.1).

Ultimately, new physical facilities (including buildings), research equipment, laboratories, and so on, would routinely become vFacilities. We hope this will amplify mission capabilities of the campus a thousand fold. While currently Centennial Campus has about 10,000 residents (faculty, researchers, students, staff), about 3 million square feet of space and some 60 on-campus industrial and government partners, without vCentennial capabilities when Centennials 1,300 acres are is fully built out, this might grow to some 30,000 residents, 200+ partners and perhaps 9 million square feet of partner space. With vCentennial we envision thousands of vPartners and capabilities that can reach a student, researcher and faculty populations world-wide.

2.4   Summary

North Carolina State University (NCSU) is embarked on an ambitious vision to change the paradigm for higher education and research by virtualizing its award-winning Centennial Campus. Centennial Campus is a small city made up of NCSU research, teaching and outreach facilities, entrepreneurs, academic entities, private firms, and government agencies. NCSU wants the ability to replicate services and functionality of this physical environment and its virtual avatars anywhere, anytime in the world using a cloud of clouds computing platform. The initial operating systems for this platform is NCSUs Virtual Computing Laboratory (VCL) technology, a powerful but low cost, and easy to use, private cloud developed over the last 8 years in collaboration with IBM. VCL now powers a considerable number of information technology offerings to students and researchers both at NCSU and world-wide.

References

1. Adair, R.J., Bayles, R.U., Comeau, L.W., Creasy, R.J.: A Virtual Machine System for the 360/40, IBM Corporation, Cambridge Scientific Center Report No. 320 (May 1966)
2. Armbrust, M., Fox, A., Griffith, R., Joseph, A.D., Katz, R., Konwinski, A., Lee, G., et al.: Above the Clouds: A Berkeley View of Cloud Computing, Electrical Engineering and Computer Sciences, University of California at Berkeley, Technical Report No. UCB/EECS-2009-28 (February 10, 2009), http://www.eecs.berkeley.edu/Pubs/TechRpts/2009/EECS-2009-28.html
3. Armbrust, M., Fox, A., Griffith, R., Joseph, A., Katz, R., Konwinski, A., Lee, G., Patterson, D., Rabkin, A., Stoica, I., Zaharia, M.: A view of cloud computing. Communications of the ACM 53(4), 50–58 (2010)
4. Bernold, L.E., Lloyd, J., Vouk, M.: Equipment Operator Training in the Age of Internet2. In: NIST SP 989, 19th International Symposium on Automation and Robotics in Construction (ISARC), Gaithersburg, Maryland, pp. 505–510 (2002)
5. Case, G.N., Vouk, M., Mackenzie, J.: Real-Time Full Motion Stereo Microscopy Over Internet. Microscopy and Microanalysis 5(suppl. 2), 368 (1999)
6. Dreher, P., Vouk, M., Sills, E., Averitt, S.: Evidence for a Cost Effective Cloud Computing Implementation Based Upon the NC State Virtual Computing Laboratory Model. In: Gentzsch, W., Grandinetti, L., Joubert, G. (eds.) Advances in Parallel Computing, High Speed and Large Scale Scientific Computing, vol. 18, pp. 236–250 (2009) ISBN 978-1-60750-073-5
7. Dreher, P., Vouk, M.: Utilizing Open Source Cloud Computing Environments to Provide Cost Effective Support for University Education and Research. In: Chao, L. (ed.) Cloud Computing for Teaching and Learning: Strategies for Design and Implementation, pp. 32–49. IGI Global (2012)
8. Garrison, G., Sanghyun, K., Wakefield, R.: Success factors for deploying cloud computing. CACM 55(9), 62–68 (2012)
9. Crnković, I., Sentilles, S., Vulgarakis, A., Chaud, M.: A Classification Framework for Software Component Models. IEEE Transactions on Software Engineering, 593–615 (September 2011)

10. Ludaescher, B., Altintas, I., Bowers, S., et al.: Scientific Process Automation and Workflow Management. In: Shoshani, A., Rotem, D. (eds.) Scientific Data management – Challenges, Technology and Deployment, ch. 13, pp. 467–507. CRC Press (2010)
11. Schaffer, H.E., Averitt, S., Hoit, M., Peeler, A., Sills, E., Vouk, M.: NCSUs Virtual Computing Lab: A Cloud Computing Solution. IEEE Computer, 94–97 (July 2009)
12. Vouk, M.: Cloud Computing - Issues, Research and Implementations. Journal of Computing and Information Technology 16(4), 235–246 (2008)
13. Vouk, M., Rindos, A., Averitt, S., Bass, J., Bugaev, M., Peeler, A., Schaffer, H., Sills, E., Stein, S., Thompson, J., Valenzisi, M.: Using VCL Technology to Implement Distributed Reconfigurable Data Centers and Computational Services for Educational Institutions. IBM Journal of Research and Development 53(4), 1–18 (2009)
14. Amazon Elastic Compute Cloud (Amazon EC2), http://aws.amazon.com/ec2/ (last accessed November 3, 2014)
15. IEEE, IEEE Launches Pioneering Cloud Computing Initiative (April 4, 2011), http://www.computer.org/portal/web/pressroom/20110404cloud
16. Internet Reactor Labs, North Carolina State University, Nunclear Engineering, http://www.ne.ncsu.edu/nrp/irl.html (last accessed November 3, 2014)
17. Mauna Kea Observatories, Hawaii, http://www.ifa.hawaii.edu/mko/
18. North Carolina State University Centennial Campus, http://centennial.ncsu.edu/history.php (last accessed November 3, 2014)
19. NIST, NIST Cloud Computing Program, http://www.nist.gov/itl/cloud/ (last accessed November 3, 2014)
20. VCL-Apache, http://vcl.apache.org (last accessed November 3, 2014)
21. Virtual Computing Laboratory, http://vcl.ncsu.edu (last accessed November 3, 2014)

Chapter 3
"Evolution, Not Revolution": iSeries – Modern Business Platform

Mariusz Kozioł

Abstract. The revolution and evolution are two only ways of progress. This rule also applies to computer architecture, computer systems, informatics an Information Technology history. Examples of both of them may be easily identified. The spectacular example of evolution is IBM i. It may be assumed that basis of successful system evolving they are system foundations. First part of paper is a short description of IBM i foundations and history.

The system, even potentially evoluable, will not be evolute if is not widely used. It should be well-known, proposed as a solution in IT structure, and used. Hence the important thing is a proper knowledge of IT specialist and IT architects. This knowledge should be initiated as a part of academic education of IT specialists. The second part of the paper is a presentation of IBM i Series introduction into academic curriculum environment in Chair of System and Computer Networks at Wrocław University of Technology.

3.1 General Considerations about History

The history shows us that there are two general ways of progress:

- evolution
- revolution

These situations can be observed at all areas, including in the field of computer and information technology. Both of them, in general, have advantages, disadvantages and characteristic features.

Mariusz Kozioł
Systems and Computer Networks, Wrocław University of Technology ,Wrocław, Poland
e-mail: mariusz.koziol@pwr.wroc.pl

© Springer International Publishing Switzerland 2015   29
R. Klempous and J. Nikodem (eds.), *Innovative Technologies in Management and Science*,
Topics in Intelligent Engineering and Informatics 10, DOI: 10.1007/978-3-319-12652-4_3

3.1.1 Revolution vs Evolution

Can be formulated several features, which may be considered as characteristics of both situations, respectively, though are not undisputable.

Revolution:

- necessery when we reached the "dead end" and there is no relatively simple way back or forward,
- revolution is usually very expensive solution,
- inevitable if we selected the wrong starting point or bad decisions during evolving process, developing,
- high level risk of making general mistakes (bad foundations, bad start). Usually we start almost "from scratch".

Evolution:

- minimized risk of general mistakes,
- little cost at a time, in terms of small steps,
- may be very slow or may not converge to the desired shape,
- must be carefully observed, it does not lead to a dead end,
- the correct decisions are difficult because evolution is dynamic process and goals may change,

It is worth noting that sometimes there are situations when, in long-term settlement, a revolution may be less expensive than slow convergent evolution.

3.2 Success Story: IBM System i

As was previously indicated, the ways of evolution and revolution can be observed also in the computers technology and Information Technology. It is worth emphasizing (as is often forgotten), that IT is a very broad area that contains a high level technologies, but also includes engines - operating systems, networking layers and hardware solutions.

Spectacular example of successful system evolution is IBM i.

The subject is large, let us consider selected key elements only.

3.2.1 Rising of AS/400

The System was introduced on June 21, 1988 as a continuation of the System/38. At start most noticeable differences were:

- added source compatibility with the System/36,
- removed capability-based addressing,
- placed in computer systems classified as rated a C2 security level

The chief architect of the System was dr Frank G. Soltis.

Fig. 3.1 Fortress Rochester
- The Inside Story of IBM
i Series. Frank Soltis wrote
in it all the assumptions,
mechanisms and principles
of the system. To this day, is
the basis for understanding
the system.

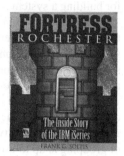

3.2.2 AS/400, OS/400

At the basis are important key assumptions of the being designed system. They were as follows.

System should:
- be optimized for applications class to which it is dedicated,
- easily utilize technology changes/improvements without impact to applications and end-user,
- be flexible and scalable with business requirements growing,
- be powerful but easy-to-use for end-user,

By design, the system was not supposed to be a universal system, but dedicated and optimized to commercial applications.

As result, the system was designed as a system for small and intermediate business, but with the sophisticated solutions that are taken from the mainframe architectures.

Most significant system foundations and features [1]:

- Technology Independance
- Object-based Design
- Single-Level Store

Technology independence is achieved thanks to introduction of Technology Independent Machine Interface (TIMI). TIMI is a kind of hardware abstraction layer. It makes we can easily utilize most hardware technology changes without rebuilding the system.

Object-based design means that there is nothing that is not an object. Object philosophy and techniques support TIMI realization and provide a strong foundation for building a system's security mechanisms.

Single-Level Store is a mechanism similar to the well-known mechanism of virtual memory, but covering whole system not just RAM. In other words: the whole system is an extension of RAM, and all objects are always in "memory" called storage (in terms of end-user) (Fig. 11.2). There are no logical disks or directories, all objects exist in "storage". It supports manipulation of physical storage elements without impact to the end-user. Single Level Store (aka Single Level Storage) induces the need for a mechanism for grouping / organizing of objects. As an ordering mechanism it is used two-level system of libraries. Libraries are universal, the key of grouping is up to user.

Due to system specialization (commercial applications) there is database embedded into operating system. This embedding supports performance and security of database. The database is DataBase 2 Universal Database (DB2 UDB). The system architecture allows the use of 1 to 32 processor sockets (one to eight core processors per socket), large amounts of memory, physical media (HDD, tape, optical media), and connecting modules. For example, current generation, based on POWER7 processor machine top model 795 can be equipped with 256 4GHz processors and 8TB RAM.

System foundations became a key for success and long evolution path.

OS/400 is a kernel of successive generations for over 20 years.

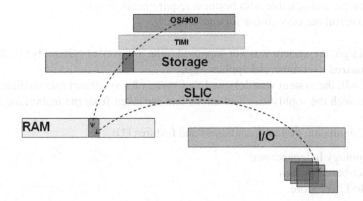

Fig. 3.2 Single-Level Store idea. SLIC (System Licensed Internal Code) is software layer providing TIMI. The SLIC software is functionally equivalent to the strict kernel of a Unix or Linux operating system. "Storage" is an abstraction, layer in which objects are addressed.

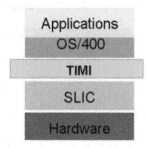

Fig. 3.3 AS/400 is a layered, library-based System. Libraries are consequence of Single Level Storage and number of objects.

3.2.3   AS/400 Evolution in Shortcuts

Thanks to its architecture and scalability AS/400 has became suit, very popular system. Following are major steps of system evolution (true and near rebranding ones) till now.

1995 -   the RISC PowerPC RS64 64-bit CPU processor replaced the 48-bit CISC processor
1999 -   added Logical Partitioning capability (i5/OS) (Fig. 11.4)
2000 -   included into eServer family as iSeries
2004 -   eServer i5 (utilization of POWER5 CPU, architecture ready for 128bit processors)
2006 -   IBM System i
2008 -   included into IBM Power Systems family (IBM i) [2]

Milestone was introduction of Logical Partitioning (LPAR).

From this point the physical resources of machine (like memory, processing time, some devices) may be divided into logical subsets called partitions. A partition becomes separated environment for operating system. Additionally it is introduced mechanism of virtualization of physical storage. Virtualized storage can be provided across partitions.There is no host Operating System, storage virtualization may be done by running in partition system. Only mechanism, called Power Hypervisor manages and guards partitions. It also supports storage virtualization by providing and maintaining virtual SCSI channels.

Is noteworthy that, unlike in other virtualization systems, here the processing time used by the partition can be closely controlled.

Thanks to Dynamic LPAR resources assigned to the partition may be changed when the partition is active, ie when OS is working. Of course OS must support

Fig. 3.4 POWER Logical Partitioning (LPAR) - True parallel, almost independent systems. No host OS. In addition there is mechanism Dynamic LPAR giving the possibility to change the partition resources without partition deactivation

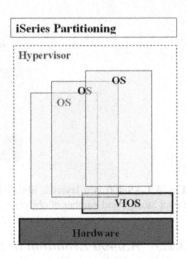

proper mechanisms (including communication with Hypervisor) for safe dynamic manipulation of resources.

Systems dedicated to run inside partitions are:

- OS/400(..),
- Advanced Interactive eXecutive (AIX - a flavour of UNIX),
- Linux (compiled for POWER processor architecture),
- Virtual Input Output System (VIOS). VIOS is specialized for storage virtualization AIX (Fig. 11.4).

Logical Partitioning allows for efficient use of server machine and the implementation of cross-system high availability solutions inside single hardware machine. It enables easily running production and test versions of systems together, without additional cost.

Other steps are often focused on expanding to various technologies, including a bundle of Internet and e-business technologies.

And finally .. current version is OS/400 IBMi 7.x (Fig. 11.5)

It includes OS/400 kernel with integrated DB2 UDB, TCP/IP stack, most known TCP/IP severs and services, Java Virtual Machines, WEB servers, application servers, PHP, SQL, LDAP, programming enironments, High Availability solutions ..and so on. Machine utilizes Logical Partitioning, cross-system and cross-machine cooperation, energy management. Systems and machines can be managed together inside defined pools. The list is very long. [3, 4].

Fig. 3.5 System version

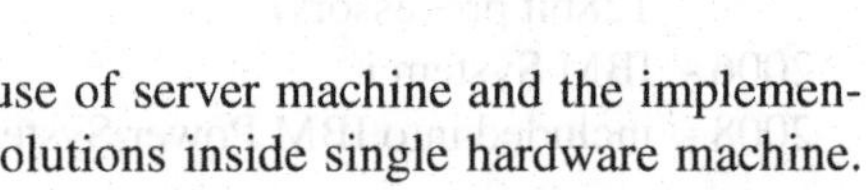

Opt	Product	Term	Feature	Description
	5770SS1	V7R1M0	5050	IBM i
	5770SS1	V7	5051	IBM i
	5770SS1	V7	5052	IBM i

3.3   iSeries on CSCN WUT

Even the best possible system is not worth much if there are no people who know him properly and is poorly utilized. Another important issue is also that IT professionals should be able to identify appropriate systems and solutions for specific need. Therefore, knowledge and basic skills in key systems should be acquired in the course of a standard educational path of IT professionals.

This was duly noted and taken into account in

Chair of Systems and Computer Networks, Wrocław Universtiy of Technology (CSCN WUT)

Later in this be presented briefly the solutions and some results.

Due to its target, the POWER technology is not widespread as other common solutions (for example x86, x64 architectures). Therefore, the beginnings were not easy.

3.3.1   2007 Academic Initiative..

Adventure with the i5 could begin thanks to activity of IBM Academic Initiative (IBM AI), section oriented to the POWER Systems . It has provided:

- access to the dedicated partition
- AI Summer School
- educational materials

Access to the dedicated partition enabled the organization of the laboratory for the students. It was possible to present the basic issues to students and they could to practice them. They could work with system via three possible interfaces: console emulator ("green screen"), Windows client application and IBM Director WEB Interface. (Fig. 11.6, 3.10) For obvious reasons, the range of possible tasks and access was limited to the level of user and programmer. From start, the important were regular meetings "Academic Initiative Summer School" (basics and advanced technologies workshops) organized in cooperation with COMMON Poland.

During the meetings there were organized labs and lectures. The lecturers were IBM workers from various countries as well as invited specialists from another companies. Very interesting was possibility of seeing iSeries machines inside - hardware technology and quality of making. Disassembling and assembling of servers were also possible. (Fig. 11.7)

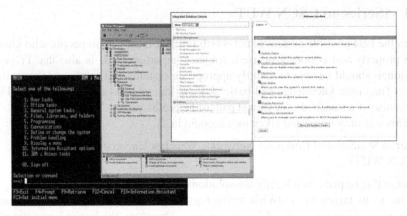

Fig. 3.6 System interaction interfaces

Fig. 3.7 AI Summer School meetings

3.3.2   2009yr Starting Own Lab..

Working on a remote partition resulted in limitations to the simple operating system issues. In 2009 CSCN decided to organize own laboratory based on 520 model of iSeries with Hardware Managemant Console.

Local server set gave a full administration and configuration possibility. The result was the ability to widening the scope of activities for students. The machine was entry level but allowed:

- presenting virtualization mechanisms
- starting number of partitions, different OSes, supervised locally.

It was doable to organize students' activity oriented to advanced configuration and organize extracurricular groups interested in administration tasks and individual projects connected with iSeries.

In 2011 year the lab is growing. Due to growing need and for technology refresh server base was extended with new, POWER7 machine, model 720.

Fig. 3.8 Locally managed several partitions

Fig. 3.9 CSCN Servers:
520, 720 and HMC

3.3.3   Curriculum Offer

Just from beginning in the CSCN curriculum appeared lectures and labs giving to the students basic knowledge about whole system, software and hardware.

The offer increased and now it includes areas such as:
- basics (general overview of the entire system, basic skills)
- programming in OS/400 environment (CL, RPG, Java..)

Fig. 3.10 CSCN Students laboratory

- introduction to the administration of the system
- network configuration and administration

Basic courses are both in polish and english languages.

3.3.4 wrkIBMi..

Very early it constituted scientific group named wrkIBMi.

The group members are students interested in IBM technologies connected with iSeries, extending curriculum provided knowledge. They must declare the area of

Fig. 3.11 wrkIBMi promotion activity

Fig. 3.12 Presentation during COMMON Poland Conference

interest and formulate the project, on which they will work on they own way. These students receive a high level of privileges in the system and necessary support.

In addition to individual work, group organizes outbound workshops, special invitation lectures, promotion activities.

One of the most interesting activity is participation in the COMMON Conference with presentation of the work results. (Fig. 3.12).

References

1. Soltis Frank, G.: Fortress Rochester: The Inside Story of the IBM iSeries. 29th Street Press, Loveland (2001)
2. IBM Archives Rochester chronology,
   `http://www-03.ibm.com/ibm/history/`
   `exhibits/rochester/rochester_chronology/`
3. IBM i For Power Systems including AS/400, iSeries, and System i,
   `http://www-03.ibm.com/systems/i/`
4. IBM Power Systems, `http://www-03.ibm.com/systems/power/`

Fig. 3.12. Presentation during COMMON technical conference

interest and formulate the projects on which they will work on their own will. Thus, students receive a high level of privilege in the system and necessary support. In addition to individual work, group organizes workshops special in various lectures, promotion activities.

One of the most interesting activity is participation in the COMMON Conference with presentation of the work results (Fig. 3.12).

References

1. Soltis Frank G.: Fortress Rochester: The Inside Story of the IBM iSeries. 29th Street Press, Loveland (2001)
2. IBM Archives Reference chronology.
   http://www-03.ibm.com/ibm/history/
   exhibits/reference/reference_technology
3. IBM i For Power Systems including AS400, iSeries, and System i.
   http://www-03.ibm.com/systems/i/
4. IBM Power Systems. http://www-03.ibm.com/systems/power/

Part II
Innovation Proposals in Management Area

Chapter 4
A Smart Road Maintenance System for Cities – An Evolutionary Approach

Hari Madduri

Abstract. One of the biggest challenges faced by cities today is maintaining their roads. Smart Road Maintenance can lead to more operational efficiency, timely response to citizens, and cost-effectiveness. Efficient and effective management and maintenance of a city's road infrastructure not only improves the quality of life for its citizens, but also makes the city more attractive for business investment. The IBM Kraków SWG Laboratory, in the cooperation with AGH University of Science & Technology, Kraków and the city of Kraków, created an innovative solution for road maintenance problems. In this paper we describe the challenges currently faced by many cities in road maintenance and how one could bring low-cost, ubiquitous technology to help cope with those challenges. This project also exemplifies the Smarter Planet/Smarter Cities initiatives being promoted by IBM worldwide.

4.1 Introduction

According to a World Bank report, for the first time in history, more than half of the world population is already living in cities [5]. By 2050, not only is the world population expected to grow from about 7 billion to over 9 billion, but also over two-thirds of that population is expected to live in cities [6]. While it is possible for brand new cities to be developed, most of the population growth and migration will happen in existing cities. This puts a considerable burden on many cities' already strained systems of transportation, roads, healthcare, utilities, etc. Thus the need for smarter management of city services is imperative. One such area that we focus on in this paper is the maintenance of road infrastructures.

Currently, many cities suffer from the problem of poorly maintained roads. There can be several reasons for that. For example, it could be insufficient maintenance budgets; or even if there are sufficient budgets, the reasons may be lack of timely identification, classification, and prioritization of maintenance problems.

Hari Madduri
IBM, 11501 Burnet Road, Austin, TX 78750, USA

© Springer International Publishing Switzerland 2015   43
R. Klempous and J. Nikodem (eds.), *Innovative Technologies in Management and Science*,
Topics in Intelligent Engineering and Informatics 10, DOI: 10.1007/978-3-319-12652-4_4

Sometimes it could also be the lack of proper information transfer and coordination among the various city departments and sub-contractors the city uses to repair the roads. In some cases, it could be just the poor quality of work performed by a sub-contractor and the lack of visibility to that at higher levels of city administration. Added to all these problems on the city's side is the citizens' perception that their municipalities don't do much for them. Often times when citizens complain about something, they have no idea if anyone is listening and acting on their complaints. Thus lack of responsiveness is also a major issue in many cities.

These problems have existed in many cities for decades. However with the advent of ubiquitous smart phones and integrated service management software, it should be possible now to address these problems much more effectively. There are already pilot projects in cities around the world exploiting smart phones to address such problems (e.g, CitySource [1], FixMyStreet [2]) . In a similar vein, our project proposes a Smart Road Maintenance System that allows smart phones (or even simple phones) to report road incidents and process them using a sophisticated service management software platform.

While what we describe in this paper falls into the general category of these referenced efforts [1, 2, 3], it differs in its breadth and focus. These referenced efforts focus on the front-end of problem reporting, while our project addresses the end-to-end solution, starting from the front-end incident reporting, through the back-end intelligent incident processing, it goes all the way to the problem resolution and other downstream activities. As our focus is more on the back-end, our work can be used to complement CitySource and FixMyStreet.

4.2 Solution Concept and System Structure

4.2.1 Concept

The concept behind our proposed solution is fairly simple. The city's road problems such as pot holes, broken water lines, fallen trees, etc can be easily reported by a citizen passing by. For example, when a citizen notices a pot hole in the road he can take a picture of the pot hole and email it to a well-known address published by the city Government. Many phones are equipped to capture GPS coordinates of the location where the picture is taken. This can be transmitted as metadata along with the picture to the city's published email address. The city receives the incident, locates it on the map and then appropriately processes the incident. This concept is illustrated in Figure 4.1.

There are a number of advantages in a system that is so simple and yet so effective:

- Due to the ubiquitous nature of cell phones today, any citizen can notice road problems and report them.

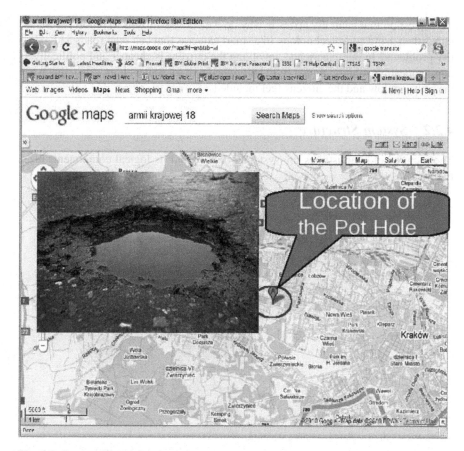

Fig. 4.1 Concept Illustration

- If someone has a cell phone with a camera, then they can take the picture and send it in as described above, but even if someone doesn't have a camera, they can call a published number and leave a recorded message describing the issue. (As we will describe later, these recordings can be processed using a speech-to-text translation software and then using text recognition to create incidents out of them).
- If someone has a smart phone, the input can be made even more structured by a dedicated mobile application downloaded by the citizen to his/her cell phone (By the way, this is the approach taken by CitySource, and others [1, 2]).
- One of the attractive aspects of this email based reporting is that the sender can be acknowledged by the city and even notified when the city acts on the reported incident.
- This ability to talk back to the reporting citizen is helpful in other ways also. For example, to send him back a reference number that he can use to track the progress of his incident. Or maybe even claim some rewards from the city, if such incentives are offered by the city.

A question that naturally arises is why do we choose citizens to report problems rather than using some technology like smart cameras that can scan roads and identify problems? In our experience, involving the citizens is cheaper and politically a more appealing solution than a fully automated one for many cities. Also, this way the costs are low and the citizens feel included.

4.2.2 System Structure

The picture below illustrates the system structure and its logical components (see Figure 4.2).

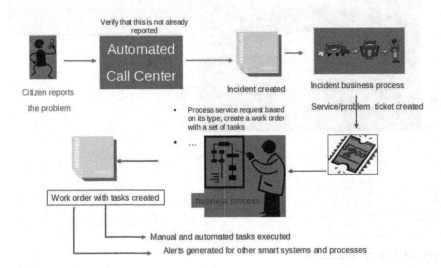

Fig. 4.2 System Structure of the Solution

The structure of our proposed solution is depicted in Figure 4.2. As can be seen, it is fairly straightforward. The citizens report problems with the roads, and the problems go through an automated call center for appropriate de-duplication (i.e., eliminate or at least reduce duplicates). An incident record is created then from the problem report. The incident is then subjected to an incident business process, where appropriate classification and prioritization is done. When it is determined that the incident requires some action, a service request is created. The service request goes through its own business process, where a determination is made on whether some work needs to be performed, and if so a work order or a set of work orders is created. These components of the solution structure and their variations are described below.

4.2.3   Component Descriptions

4.2.3.1   Input

There are multiple ways in which incidents can be reported. The simplest being someone taking a picture of a pothole (or for that matter any road problem) and sending it to the published city email address (e.g., e-mail: `road-problems@yourCity.com`). There are a number of other ways:

- Simple phone call to a published city phone number and leaving a voice message
- Web interface, where a form can be filled out from a personal computer, for example.
- Web interface, accessed via smart phone
- A dedicated mobile application on a smart phone (e.g, iPhone, Android or Blackberry)
- An SMS message, etc.

4.2.3.2   Automated Call Center

All calls, whether delivered through phone messages or emails, or web-based forms, need to be processed by a call center. While this in theory can be a manual call center, in practice it needs a fair amount automation, to cope with the volume of incidents generated as more and more citizens report problems eagerly. One of the main functions of such a call center is validation and removal of duplicates.

4.2.3.3   Incidents and Incident Business Process

Once a reported incident is validated (i.e., not a duplicate), it must be registered and subjected to a process of classification and prioritization. Some city administrations may want certain class of events to be physically verified by one of their city personnel. For example, some city Governments may choose not to trust any incident unless it is verified by a designated department staff. This becomes part of the business process. Some incidents may be classified as dangerous (e.g., a puddle of water covering a big pot hole) and some urgent attention is needed even before someone repairs it. These actions are also triggered based on the classification. Sometimes incidents may be combined into a larger one.

4.2.3.4   Service Request (Ticket) Creation and Process

Once an incident is verified, classified and prioritized, someone needs to act on it. This action is requested via a service request (also called service ticket). A given incident may result in one or more service requests. Service requests are processed using their own business process and fulfillment of service requests results in work orders being created.

4.2.3.5 Work Orders and Events

A work order is another business object, much like incidents and service requests. It is subjected to its own business process or workflow logic. Work orders in general contain tasks that need to be performed. As a side effect of work order execution, events (or alerts) can be generated to communicate with other systems (e.g., repairing a pot hole may require co-ordination with the water department to turn off/on water lines). Some work orders may have child work orders that are delegated to city's sub-contractors. The work order management system ensures that all work, whether performed by the city or sub-contractors is properly tracked and coordinated. Keeping a history of work orders is also helpful in making warranty claims on sub-contractors (and in the long run developing a vendor rating system.)

4.3 Making Things Smart

The process of reporting and handling road incidents or service requests may look fairly straightforward and routine. However, the use of appropriate technology makes the process more efficient, scalable and affordable. This is what makes it smart, compared to the current way most cities approach road maintenance today. In general, making any city solution smarter involves leverage of technology to reduce cost, increase speed, and providing better services.

In what follows we describe some example 'smarts' that can be put into a road maintenance solution. Some of these have already been implemented in our pilot project and some others are future directions. (The current status is summarized in the Results section.)

4.3.1 Faster and Mostly Automated Problem Reporting

Smart phone technology can be leveraged to quickly and automatically report problems. For example, with a dedicated smart phone application, one can take a picture, create an incident attaching the picture and its location meta-data, and send it to the city; all in a matter of a few clicks.

4.3.2 Automated Localization and De-duplication

The location information, for example the GPS co-ordinates of where the picture was taken, are sent as meta-data and using the APIs of a mapping service the incident is located on a map (see Figure 4.3).

We can also eliminate duplicates by comparing the incident meta-data (for example, by seeing whether two incidents have very close GPS coordinates).

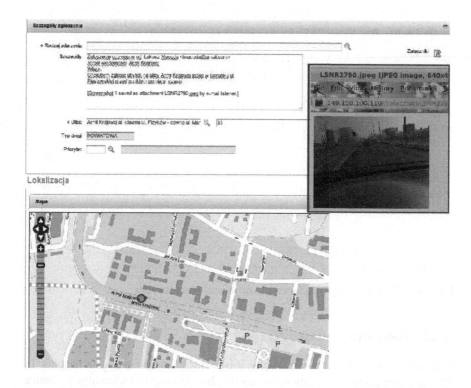

Fig. 4.3 Locating the incident on the map using GPS coordinates in the metadata

4.3.3   Automated Processing, Classification, and Prioritization

The incident records created as above contain structured data and hence can be processed by software very well. For example, based on the location of the incident and the citizens marking of severity level, it can be classified and prioritized automatically.

4.3.4   Rule-Based Processing of Incidents, Service Requests and Work Orders

In general, depending on the service management software used, it might be possible to process all objects (i.e, problems, incidents, service requests, and work orders) based on business rules. [Figure 4.4 depicts the workflows we implemented in our City of Kraków pilot project. The tool used was IBMs Tivoli Service Request Manager. The thunderbolt icons on arrows indicate actions (snippets of Java code) that get called when those state transitions occur.]

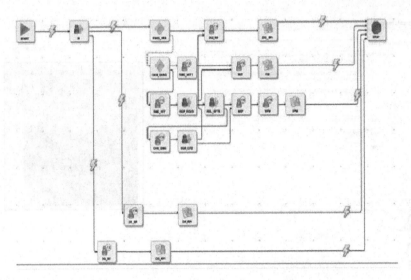

Fig. 4.4 Incident handling Workflows can incorporate business rules

4.3.5 Analytics

Since all data is gathered automatically, it is easier to analyze and gain insights. In general making any city solution smarter involves leveraging technology to reduce cost, increase speed, and providing better services.

4.4 Results and Project Status

4.4.1 Results

Most of the design has been implemented in a pilot project done for the City of Kraków (Poland). Here are some details of the pilot implementation:

- Three input methods of creating road incidents were implemented: photos from phones, regular email, and website-based form.
- IBM Tivoli Service Request Manager (TSRM) product was used to implement the service desk (call center). Using its email-listener interface, incident records were automatically created from received emails. Both smart-phone based picture emails and regular emails came in this way. The web application was fairly straightforward to build as a servlet that talked to TSRM to create an incident.
- The image meta-data was used in conjunction with Google^{TM} maps (as the map provider) to locate the incidents on the city map. (In a later implementation we have supported OpenStreet Maps as yet another map provider. The architecture is flexible enough to work with different map providers.)

- The incident data was also exploited to automatically identify the resource the incident applies to (e.g., streets were treated as the resources/assets to which incidents apply). In the pilot implementation, we have input all city streets (street segments) as assets into IBM Tivoli Service Request Manager. Based on the GPS coordinates of the incident, we were able to automatically identify the road segment that the incident was on. This also helps to automatically assign the incident to the City employee (like a dispatcher or supervisor) responsible for that road segment.
- De-duplication, classification, prioritization, etc. were done manually by the dispatcher (a human role) assigned to the incident. (in a later implementation, we have done some automated de-duplication by exploiting the proximity of GPS locations. I.e., they don't have to be identical but if they are close enough, then they are treated as duplicates.)
- Analytics: In the pilot implementation we only produced some simple analytic reports that calculated the average duration of problem resolution. However, there is a lot more potential here. As one can observe, the system accumulates a wealth of information with timestamps as the problems are reported, incidents tickets created, service requests created, work orders generated, passed on to subcontractors, and tracked. Overtime this can be exploited to identify bottlenecks in the system and reduce the overall cycle time. This leads to better citizen satisfaction besides optimizing the use of limited city resources.

The resulting pilot system was deployed for use by the city construction company (ZIKIT A City of Kraków Company responsible for Road Maintenance) and was successfully used for a pilot period of about two months. Initially, the users were restricted to be ZIKIT personnel only, but once it was extended to ordinary citizens via publicity on ZIKIT's website, several citizens participated and the number of incidents went up significantly.

In the pilot project the final part of generating work orders and tracking their progress with the repair companies was not implemented. (This is currently done by ZIKIT using their internal work management system. Changing that was beyond the scope of pilot). As a result there are no measurements on the change in citizens' satisfaction due to the pilot. We did however receive several favorable comments that the citizens and the press liked the pilot system.

(As a follow on to this project, we are pursuing an extended pilot in 2013 that is implementing the sub-contractor functionality as well.)

4.4.2   Status

At the time of the original writing (mid 2012), we started moving past the pilot in two parallel directions. One is to implement an extended pilot for the city of Kraków with more functionality and scope of citizen coverage. The other, working jointly with AGH University, is to push the Research and Development direction to experiment with more 'smarts' in de-duplication,

4.5 Comparison to Related Work

4.5.1 Similar Systems

There have seen several efforts similar to ours. For example, CitySourced, a California-based company, offers a smart phone application that can be downloaded by citizens from their website and used to report civic issues to their city Governments. All the reported events accumulate at CitySourced server. City Governments can set up a business arrangement with CitySourced to get incidents relevant to their city and then act upon them. The interface for citizens is fairly simple to report civic incidents like pot holes, broken water mains, graffiti, street fights, illegal dumping, etc. All these incidents are also shown on a regional map on CitySourced's web site.

A nice feature of CitySourced is easy reporting and intuitive depiction of incidents on a regional map so citizens know where all the reported issues are. Like our pilot system, CitySourced smart phone application can also take a picture of the incident and report it to the CitySourced.com website with the picture and some additional information. Further the reported incidents follow an industry standard (Open311 [7] events), making it easy for other systems to consume them. The citizen can also look up the status of any incident (regardless of who reported).

Another very appealing feature of CitySourced is that it is a SAAS (software as a service) model where a city Government doesn't have to buy or implement any software. It simply subscribes to civic events from the service and receives them in Open311 format.

In comparison, our solution also has the simplicity of reporting city problems using a cell phone, and we make it even simpler than them. For example, we don't require a smart phone application, though we could use it, if one exists. We can let even ordinary cell phones (i.e., not so smart) take pictures and report problems. Our goal was to minimize the barriers to adoption. If people don't need any special software then the solution is easier to adopt.

Another system similar to CitySourced is FixMyStreet which seems to be widely used in the UK. This system primarily uses the web interface for citizens to report problems. The web interface is easy to use and it presents a map for citizens to locate their problem on the map. It also shows already-reported problems in the vicinity of the problem one is about to report. This allows for some de-deplication and location work to be avoided as the citizen entering the report will avoid reporting duplicate incidents.

FixMyStreet also has reports by area (city) where they can show a summary of total problems reported, newly reported, old and fixed, and not fixed, etc. This gives a good view to citizens how effectively their municipalities are dealing with their reported problems.

There are other systems similar to the above two. For example Naprawmy [3] in Poland is a community project that is being offered on a voluntary basis to cities.

All of the above solutions focus on the input subsystem of an overall city infrastructure maintenance system. Also these systems are not limited to road problems, but allow any kind of city problems to be reported. In contrast, our approach addresses the end-to-end solution, with more focus on the back-end processing of reported problems. All the other compared systems don't do much backend work.

End-to-end solution: It is probably useful here to distinguish an end-to-end solution for road maintenance from other solutions that only address a subset of aspects. At one end of the solution is the citizen that notices a problem, reports it and expects a response/resolution within a reasonable amount of time. As his/her problem enters the system and as it gets transformed into an incident, then as a service request, then as a work order and then as a project (or a set of tasks), we need to have appropriate management systems (e.g., including classification, prioritization, deduplication, etc.) for each of these transactions. Further, for continuous improvement these systems also need to generate measurements and allow for tuning. At the far end of the solution, after the problem is resolved, the originator needs to be informed and a set of metrics need to be collected for evaluating the efficacy of personnel/contractors doing the work. Now, an end-to-end solution needs to focus on all stages, not merely reporting the problem via a cell phone or merely generating metrics on how long the resolution took. While the end-to-end solution must address all stages, it is not necessary for a single product to address them all. The solution can be composed of multiple products focusing on different aspects/stages.

Besides being an end-to-end solution, ours also focuses on flexibility and pluggability. Thus it is possible for our solution components to be replaced by other products that fulfill specific functions. For example, we could replace or augment our solution with someone else's mobile application that produces Open 311 events to report problems.

All the above compared systems can be complementary to our solution. While all of them can use some analytics based on incident data, we have the ability to gather more data and apply more analytics on the backend, because of our focus on backend processing.

To summarize, our system has the following differences:

- We chose a low barrier for adoption by citizens by providing multiple and simpler ways of reporting
- We provide an end-to-end solution, going all the way from reporting problems to analyzing them to classifying/prioritizing them, to finally turning them into work orders and tracking them.
- Due to the end-to-end focus, we have more ability to automatically gather data and leverage data analytics for better insights and more automated means of coping with scale.
- Architecturally, we have pluggability to replace components with best of breed implementations.

4.5.2 Related Work

The work we have undertaken at IBM is one of several projects within IBM under the umbrella of Smarter Planet/Smarter Cities solutions. IBM has proposed a strategic architecture and delivered a platform upon which Smarter City Solutions can be built. This is called Integrated Operations Center (IOC for short). It is an integrated offering of several IBM software group products that provides the following capabilities:

- Provides a unified view across all city agencies
- Allows supervisors to monitor and manage a range of services
- Enables agencies to respond rapidly to critical events
- Delivers situation awareness and reporting
- Streamlines management of resources and critical events
- Integrates with open-standards connection points to existing and future systems

IOC quickly allows the user to get an understanding of what, who, why and where attributes of an issue or potential issue.

The following graphic expresses the Smarter Cities Operations vision of IOC (see Figure 4.5).

Smarter Cities Operations

Fig. 4.5 IOC's vision to optimize city-wide operational systems

The smart city solution presented in this paper becomes a particular subsystem (city's road infrastructure) solution within the larger framework of IOC-based smart city system. There can be several such subsystems, and IOC facilitates integration and communication among them. In particular, appropriate events from one system to another are communicated via events built using an open protocol (Common Alert Protocol or CAP events). At a higher level IOC facilitates filtering, aggregation and analysis of data to enable city-wide visibility and optimization across city's subsystems.

4.6   Conclusion and Future Work

In this short paper, we have described a very simple and yet pragmatic approach to report and resolve cities' road problems. It relies on using ubiquitous technology and involving citizens to report problems. It leverages powerful service management software that has hitherto been applied in other industries (e.g., IT, manufacturing, and other heavy industries) for public service. Many of the solution ideas expressed here have been successfully implemented in a pilot project for the City of Kraków.

We have taken an evolutionary approach to the problem in the sense of introducing changes gradually and realizing commensurate benefits along the way. For example, we didn't require the phones to be smart phones and run our special software, we didn't require a city change its business process completely, or change the way it interacts with its subcontractors and so on. This gradual change is very important for cities to change their culture and embrace a new technology.

There are many directions for future work in this area:

One is certainly making the solution operational to full cities rather than pilot areas. This is the usual process of hardening the software and the business processes to be robust enough to meet the scale challenges.

A second direction is to put more 'smarts' into the solution components. In particular, de-duplication of incidents by exploiting image recognition and text recognition.

A third direction is to integrate the solution into a broader industry framework like IBM IOC, so that the solution can emit/consume alerts/events to & from other subsystems. This allows the road maintenance system to contribute to overall optimization of city operations.

A fourth direction is to exploit data analytics to make the solution components like Automated Call Center, Incident/Service request/Work order business processes to be more adaptive. This provides the obvious benefit of continuous improvement to the solution described in this paper.

Acknowledgements. The author would like to acknowledge the many people that contributed to this project: AGH University professors (Prof. Zieliński and others), research staff and students, the City of Kraków executives, the executives and staff of ZIKIT, and fellow IBMers Marcin Kalas, Łukasz Macuda, Robert Lizniewicz, Dominik Najder, Marek Grochowski and several others (too many to mention by name) that contributed and are continuing to contribute to this project.

References

1. CitySourced, http://www.citysourced.com/default.aspx
2. FixMyStreet, http://www.fixmystreet.com/
3. Naprawmy To ("Let's Fix This" in Polish)
   http://www.krakow.naprawmyto.pl

4. IOC (Intelligent Operations Center),
   `http://pic.dhe.ibm.com/infocenter/cities/v1r0m0/`
   `index.jsp?topic=%2Fcom.ibm.iicoc.doc%2Fic-homepage.html`
5. World Bank Report on Urbanization, `http://go.worldbank.org/V8UGUWCWK0`
6. United Nations, Department of Economic and Social Affairs, Population Division,
   `http://esa.un.org/unpd/wpp/Analytical-Figures/htm/fig_2.htm`
7. A collaborative model and open standard for civic issue tracking,
   `http://open311.org/`

Chapter 5
Cloud IT as a Base for Virtual Internship

Jerzy Kotowski and Mariusz Ochla

Abstract. Summer 2010, IBM Poland and Wrocław University of Technology introduced the first University Cloud Computing Centre in Poland. Additionally a Multipurpose Cloud Computing Centre has been created with the University of Technology being the first academia worldwide in this initiative. Main objectives of the Centre are: usage of the Cloud for education, usage of the Cloud for a Remote Educational Internship program, promotion of research on Cloud technologies and local and international collaboration with other faculties and business partners in cloud services.

A set of tools has been created in order to support the Centre and its mission. The main part is a private cloud platform for education. The provisioning time has been optimized for classrooms and virtual classrooms that is essential especially for technical workshops, instructor led demos and self-paced education.

A dedicated portal for Remote Educational Internship has been built on top of the cloud platform together with a set of activities and personalized tasks for the interns to build corporate skills that are highly demanded on the job market.

5.1 Introduction

Remote Educational Internship program has been built as a front-end application using cloud platform as a back-end. It has been created based on an observation that IT consultants, especially in big corporations, needs to work remotely, using resources in a cloud, working in international teams on complex technical and business projects. An internship should give the company a chance to work with students, to evaluate their readiness to be self-sufficient in modern virtual projects.

Jerzy Kotowski
Wrocław University of Technology, 27 Wybrzeże Wyspiańskiego Str.,
50-370 Wrocław, Poland
e-mail: jerzy.kotowski@pwr.edu.pl

Mariusz Ochla
IBM Poland Sp. z o.o., ul. 1-sierpnia 8 bud. A, 02-134 Warszawa, Poland
e-mail: mariusz_ochla@pl.ibm.com

© Springer International Publishing Switzerland 2015
R. Klempous and J. Nikodem (eds.), *Innovative Technologies in Management and Science*,
Topics in Intelligent Engineering and Informatics 10, DOI: 10.1007/978-3-319-12652-4_5

It should also give participants a valid feedback on possible gaps that their future corporate employers may easily identify.

The Remote Educational Internship, being one of the initiatives of Multipurpose Cloud Centre, is a portal through which participants receive individual tasks that completion is controlled and monitored by the system. Although the tasks are personalized for individual intern, however the exercises are always grouped in 3 phases: Education, Projects and Team Work.

First phase - automated, self-paced education in cloud is usually the longest part of the internship. Participants receive a vast set of materials on different technologies from which they are expected to choose 2 or 3 they would need in the next phases of the internship.

During the second phase the students are expected to understand IBM's Smarter Planet concept and get familiar with already existing projects and creation of an idea of a project in which the technical skills gained in the first phase may be applied in a real life.

Third phase - is a set of team activities that require a lot of maturity and well developed interpersonal skills. It is a technical or business oriented set of exercises run in a virtual team of students from different departments and specialties. Technical students are expected to create a new virtual image containing IBM software necessary to support one of the projects presented at the second phase of the internship, while business students are usually asked to create an executive presentation on the chosen project.

A list of ten training modules has been offered to interns to choose from. The students are expected to select and complete the modules that will be the most useful to complete all other exercises and task during the internship. The portal also offers a free choice for reservation of cloud resources so that students may easily decide when they should complete the education part.

The duration of the internship was one month, and all course modules have been conducted in English. Wrocław University of Technology's organized the courses, and IBM Poland has run the teaching modules based on IBM Tivoli Software and cloud expertize.

The Centre has the capacity to offer fifteen hundred internships each year. During its first summer, around four hundred students participated. Upon completion, graduates receive an industry recognized certificate awarded by the IBM Toronto Lab attesting to their mastery of database skills. The top graduates of these internships have been nominated to a scholarship program at an IBM Research labs.

In the following document, there will be presented in detail the concept of Cloud Computing, requirements concerning organization of internship at Polish Universities as a result of Bologna Agreement and such solution introduced successfully in The Wrocław University of Technology.

The main idea the Remote Educational Internship lies in giving the students an access to virtual machines for a technical, software oriented, exercises and a guidance for the educational, project-oriented and team work parts provided by the portal. All the resources have been activated for every person in cloud, according to planned and agreed schedule.

This chapter is a regular continuation of the two previous chapters, arranged by Mr. Oded Cohn and Mr. Mladen A. Vouk. In the next sections of it, there will be described the basics of Cloud Computing, the virtual machines technology and practical application of that instrument, which in the past two years, helped hundreds of students not only from the Faculty of Electronics of the University of Technology to accomplish the requirement of completing the internship during their studies. As the summary, there will be shown the plans of future cooperation between PWr and IBM.

5.2   Fundamentals of Cloud Computing

Cloud computing is a category of computing solutions in which a technology and/or service lets users access computing resources on demand, as needed, whether the resources are physical or virtual, dedicated, or shared, and no matter how they are accessed (via a direct connection, LAN, WAN, or the Internet). The cloud is often characterized by self-service interfaces that let customers acquire resources when needed as long as needed. Cloud is also the concept behind an approach to building IT services that takes advantage of the growing power of servers and virtualization technologies.

Cloud computing's importance rests in the cloud's potential to save investment costs in infrastructure, to save time in application development and deployment, and to save resource allocation overhead.

The flexibility of cloud computing is a function of the allocation of resources on demand. This facilitates the use of the system's cumulative resources, negating the need to assign specific hardware to a task. Before cloud computing, websites and server-based applications were executed on a specific system. With the advent of cloud computing, resources are used as an aggregated virtual computer. This amalgamated configuration provides an environment where applications execute independently without regard for any particular configuration [13].

The cloud computing model is comprised of a front end and a back end. These two elements are connected through a network, in most cases the Internet. The front end is the vehicle by which the user interacts with the system; the back end is the cloud itself. The front end is composed of a client computer, or the computer network of an enterprise, and the applications used to access the cloud. The back end provides the applications, computers, servers, and data storage that creates the cloud of services.

Typical benefits of managing in the cloud are: reduced cost, increased storage and flexibility. Experts assume that the challenges of the cloud computing are increasing of data protection, growing data recovery and availability or the growing management capabilities. Due to these benefits, the use of High Performance Computing infrastructure to run business and consumer based IT applications has increased rapidly during the last few years. However, clouds are essentially data centers that require high energy usage to maintain operation [1]. High energy usage is undesirable since it results in high energy cost. For a data center, the energy cost is a significant component of its operating and up-front costs. Therefore, cloud providers want to increase their profit by reducing their energy cost in different ways. In [14, 8, 13] authors show that this problem of minimizing energy consumption maps to the 2-dimensional bin-packing problem. In the modern literature papers related to this idea one may find really a lot. See [1], for example.

5.3 Virtual Machine

A virtual machine (VM) is a "completely isolated guest operating system installation within a normal host operating system". Modern virtual machines are implemented with either software emulation or hardware virtualization. In most cases, both are implemented together.

A virtual machine is a software implementation of a machine (i.e. a computer) that executes programs like a physical machine. Virtual machines are separated into two major categories, based on their use and degree of correspondence to any real machine. A system virtual machine provides a complete system platform which supports the execution of a complete operating system. An essential characteristic of a virtual machine is that the software running inside is limited to the resources and abstractions provided by the virtual machine—it cannot break out of its virtual environment.

At the moment, there exists hundreds of application that support cooperation between server and user PC, which are being installed by user. Among them, we can distinguish such as: VMware, VM from IBM, etc.

5.4 Deployment Models

Cloud platform services computing platform and solution stack as a service, often consuming cloud infrastructure and sustaining cloud applications. It facilitates deployment of applications without the cost and complexity of buying and managing the underlying hardware and software layers. For this moment at least six deployments models of the cloud computing are available. The most popular and useful deployment models presently are as follows: *a public cloud, a community cloud, a hybrid cloud* and *a private cloud*.

The main area of our interest is to built at the Wrocław University of Technology the public cloud that will provide educational software to K-12 schools on the Low Silesia district. For this moment we assume that necessary back end elements of the cloud will the computers working at our University.

Existing meta-heuristic algorithms are based on ideas found in the paradigm of natural or artificial phenomena. For our purposes we reworked an optimization procedures based on the idea of Genetic Algorithm and Harmony Search Algorithm

[14, 8]. Genetic Algorithms are adaptive heuristic search algorithm premised on the evolutionary ideas of natural selection and genetic. The basic concept of Genetic Algorithm is designed to simulate processes in natural system necessary for evolution. As such it represents an intelligent exploitation of a random search within a defined search space to solve a problem. The newest method in the area of the meta-heuristic algorithms is the Harmony Search Algorithm (HSA). It was conceptualized from the musical process of searching for a perfect state of harmony, such as jazz improvisation. Although the HSA is a comparatively simple method, it has been successfully applied to various optimization problems. Both approaches need for the very beginning an appropriate coding procedures to conduct necessary operators easy.

5.4.1 Problem Description

Given:

- The number of virtual machines. Each of them is described by needed/required resources (CPU performance, memory). Required working hours are also known. It is assumed that these requirements may be different for each machine.
- There is a cloud computing system consisting of multiprocessor computers (host machines) with defined parameters.

Assumptions:

- Each virtual machine must be deployed on a single host. It is not allow for a virtual machine to use resources from several hosts.
- The hardware resources of at least one host machines are larger than those required by the virtual machine.
- Any VM has a given priority. It is a real positive number.
- The most important parameter for the VM is its performance. The minimal feasible performance is known as like as its advisable value.
- Working hours for each VM are known.

Task:
   Deploy virtual machines on the particular host.

5.4.2 Problem Analysis

We may study two versions of the problem presented above.

- *Version I:* all parameters of the virtual machines are random numbers. Virtual machines (VM) form the stochastic process. This assumption leads to the management problem in the real time.
- *Version II:* all parameters of the virtual machines are known previously. This assumption leads to the cutting/packing problem.

One may show that this problem may be easy transformed to the Strip Packing Problem. The Strip Packing Problem or Strip Cutting Problem (SCP) is formulated as follows: to pack (or cut) a set of small rectangles (pieces) into a bin (strip) of fixed width but unlimited length. The aim is to minimize the length to which the bin is filled.

This problem is classified as 2/V/O/R. It was shown that SCP is NP-complete since it is a generalization of the cutting stock problem, which in turn is a generalization of the famous knapsack problem.

5.4.3   The Method and the Algorithm

The problem is NP-complete. More, in practice the total number of VM goes to hundreds. Decisions variables are of integer and real kind. In the problem occur many additional restrictions. We must also remember about the time consuming procedures (fitness function, decision variables coding and decoding). The final conclusion may be only one: a meta-heuristic approach should be used. Namely we decided to put into the motion the newer version of Harmony Search Algorithm (HSA) named Improved HSA (IHSA).

Classic HSA was recently developed in an analogy with music improvisation process where music players improvise the pitches of their instruments to obtain better harmony. Algorithm scheme contains the following steps:

1. Initialize the problem and algorithm parameters.
2. Initialize the harmony memory.
3. Improvise a new harmony.
4. Update the harmony memory.
5. Check the stopping criterion.

HSA uses the following parameters: *HMS* - The harmony memory size, *HMCR* - harmony memory considering rate, *PAR* - pitch adjusting rate, *bw* - an arbitrary distance (bandwidth) and *NI* - the number of improvisations (for stopping criterion).

The traditional HSA algorithm uses fixed value for both *PAR* and *bw*. In the HSA method *PAR* and *bw* values were adjusted in initialization step (Step 1) and cannot be changed during new generations. The key difference between IHSA and traditional HSA method is in the way of adjusting *PAR* and *bw*. Small *bw* values in final generations increase the fine-tuning of solution vectors, but in early generations *bw* must take a bigger value to enforce the algorithm to increase the diversity of solution vectors. Large *PAR* values with small *bw* values usually cause the improvement of best solutions in final generations which algorithm converged to optimal solution vector. Usually, this parameters changes during the calculation process due to the rules presented in (5.1)-(5.2).

$$PAR(gn) = PAR_{min} + \frac{PAR_{max} - PAR_{min}}{NI} gn \qquad (5.1)$$

$$bw(gn) = bw_{max} e^{c \cdot gn} \qquad (5.2)$$

where c in (5.2) is given as:

$$c = \frac{ln\left(\frac{bw_{min}}{bw_{max}}\right)}{NI} \qquad (5.3)$$

In (5.1)-(8.2) gn stands for the generation number.

5.5   One Hundred Years of Business Machinery Named IBM

The company IBM was born in 1889 and it is one of the first IT concerns in the world. It hires about 400 thousand of employees in departments in 170 countries. The corporation conduct projects in almost every field of informatics market, beginning with computers and software, ending with nanotechnology and applying their own innovations in cooperation with many public organizations. See Fig. 5.1.

The company that became the global behemoth IBM was founded on 16 June 1911. To celebrate its centenary, we look at some of the landmarks of business technology - and the company behind it. It is almost impossible to enumerate all milestones that had a significant influence on the development our civilization. The set of the best known is presented below [9, 11, 12].

1928	IBM introduces rectangular, 80-column punch card design, which lasted nearly 40 years and and still contributed 20% of revenues in the mid-1950s;
1931	Type 600 Multiplying Punch introduced. It could multiply two numbers entered on a punched card and return the answer in another space;
1934	IBM 801 Bank Proof machines could list, separate, endorse and record total of cheques;
1946	IBM shifts from mechanical to electronic computing with the 603 Electronic Multiplier, which boasted 300 vacuum tubes;
1949	The Model 407 Accounting Machine is introduced;
1952	IBM ships its 701 Defense Calculator or Electronic Data Processing Machine with Type 726 magnetic tape recorder and reader;
1957	Fortran, the first high level software language, is developed by IBMer John Backus;
1961	IBM's Selectric typewriter introduces the print "golfball", which sets new benchmarks for speed and quality;
1964	Launch of the IBM 360 Series spearheads IBM's drive into general business computing market;
1971	IBM engineers developed the 3330 disk drive unit, the first "floppy";

1973   First dedicated electronic point of sale products launched, IBM 3650 and 3660 Store Systems;

1981   IBM 5150 Personal Computer (PC) launched; a twin-disk drive model with 16Kb of RAM running Microsoft DOS, VisiCalc and EasyWriter on an Intel 8088 processor could be yours for just $1,565;

1995   IBM extols network-centric e-business computing and forecasts it would end the PC's reign at the centre of the computing universe;

2000   IBM commits $1bn to support the open source Linux operating system;

2008   IBM "Roadrunner" machine at Los Alamos National Labs runs more than a quadrillion calculations per second - known by experts as a "Petaflop". A hybrid supercomputer using two different processor architectures, Roadrunner is twice as energy-efficient as the next fastest computer.

Presently, the most important initiative of IBM is "Smarter Planet" - the idea consisting of many different areas, among others: energy, traffic, food, education, health, public security and more. Its main assumption is an introduction of the newest technologies in order to improve and facilitate work of number of public sections as well as to reduce time of waiting for its results. For more details see the previous chapters and [3, 10, 14, 15].

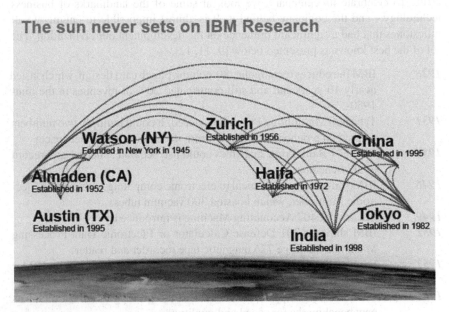

Fig. 5.1 The sun never sets on IBM Research

5.6   The History of Contacts between Wrocław University of Technology and IBM Corporation

The first important agreement between PWr and IBM was signed in 2007. From that moment the cooperation began to develop fruitfuly. 30 September 2009, in the head office of Polish Information and Foreign Investment Agency in Warsaw, there was signed a contract between IBM Poland and Polish government, concerning the location of the new IBM Integrated Delivery Center in Wrocław.

The new Center is supposed to offer IT services for the clients of IBM that have their seats in Europe. During the ceremony of signing the contract, the representatives of the Polish government and the directors of IBM emphasized many times the meaning of academic support, especially the University of Technology in Wrocław, while deciding about the location of IBM Center. It was assumed then, that at PWr, there would be created special team aimed for the cooperation with IBM. The University started the important obligations in the field of postgraduate training for employees and education for future IBM staff.

5.7   Multipurpose Cloud Center (mc2)

In May 2010, after creation of University Competence Center on Cloud and the development of the cooperation in many areas related to Cloud, called Multipurpose Cloud Center (mc2), the Faculty of Electronics and IBM Corporation introduced an innovative internship program for students of PWr. Apart from its educational character in the informatics area and a set of practical tasks, it offers a flexibility to work from any remote location in any working hours (task work). However the most important aspect of the program is that it offers the students the opportunity to prove themselves and demonstrate their creativity.

The students from the 6th semester of Bachelor degree from any faculties and departments of the University has been invited to participate in the program. According to Polish law, the completion of a professional internship is mandatory to graduate of Bachelor studies in all Polish technical universities.

The duration of the internship has been set up to 4 weeks in the summer 2010 and the program ran over three months of summer holidays: from July to September. It had a form compatible to IBM initiative, called Multipurpose Cloud Center (mc2) and similar to ESI (Educational Students Internship). In every period, almost 300 hundred students could have taken part in the event and every ten of them had been looked after by one mentor from IBM. Apart from that, the University of Technology in Wrocław selected several tutors for every 30 of trainees.

27 July 2010 IBM Poland and the Wrocław University of Technology announced signing an agreement concerning the first Multipurpose Cloud Computing Center in Poland.

Presently, there are 10 modules divided by different scientific areas in the Cloud, from which trainees may choose the most interesting for them. The system is flexible and allows a continuous growth of the repository in future.

One of the initial difficulties the students need to face during the internship is a conscious choice of their learning path and a proper selection of materials that would help to complete all other tasks during the program. The students, usually divided into groups, may have choose a different combination of workshops, adjusted to the particular group's profile or interests of its members.

All essential information about internship and the available materials is provided through a web site. The portal built on top of cloud engine allows: to reserve application, to check the schedule, notes, and attached materials as well as to obtain current information concerning internship and tasks that should be executed. There is also planned to create a chat module for a better help-desk services. Students could also be able to discuss with their mentors in the real time and keep conversation history in the same place. Such system facilitates internship management from the level of usual web browser.

The actions like creating user account, rating their work, checking the schedule and statistics are possible from an administration panel. The employees of the University might have the access to the various numbers of workshops using voucher system.

The newest environment for internship management will enable using composite systems consisting of several virtual machines with the requirement of constant network communication. Such solution will give the students possibility to implement multiple tasks based on network services. In the consequence, there will be possible to prepare advanced topologies and popularize them via Cloud Technology. See Fig. 5.2.

5.8 Events

During 1,5 year cooperation between IBM and PWr, there are many noteworthy events. Some of them, which took place just after first edition of internship, are presented below.

29.07.2010 Information about IBM and WrUT in media: „IBM Opens the First Multipurpose Cloud Computing Center in Poland". See Fig. 5.5.

02.09.2010 An invitation of Wrocław University of Technology to IBM Cloud Academy. See Fig. 5.3. The present members of this body are presented of Fig. 5.4.

08.09.2010 An application of Faculty of Electronics to the Rector of PWr concerning creating a team called Cloud Committee

07.12.2010 CIRCULAR LETTER from the Rector of PWr concerning cooperation with IBM Cloud Academy

10.10.2010–now Participation in every-month online conferences of all IBM Cloud Academy members with the help of Lotus Live Technology

10.03.2011 Presentation of PWr on the online conference

Fig. 5.2 Hot summer at the Wrocław University of Technology

05.09.2011-28.10.2011 Second edition of internship mc2 with the participation of 115 students. Organization of the contest for a team project, with the final 6 groups consisting of 4-5 members. The best team visited IBM Haifa Research Laboratory in Israel in the first week of December 2011 as a result of host's invitation

01.10.2011 Nomination of mc2 technology by the European Committee to the group of the best projects in the IT field in 2010

30.10.11-06.11.201 Visit of the representatives of PWr in IBM Haifa Research Laboratory

03.04.2012 Common organization of science conference „World of Innovation".

5.9   Summary and Future Plans

In addition to the Multipurpose Cloud Computing Centre, IBM has invited Wrocław University of Technology to join its two other cloud initiatives. The first is IBM's Academic Skills Cloud in which twenty (Q4 2010) academic units throughout the world collaborate to form a large educational Cloud project. The second is an association of elite higher education institutions working in the Cloud Computing technologies called the IBM Cloud Academy.

IBM.

Dr. Jerzy Kotowski:

IBM is pleased to welcome Wroclaw University of Technology to the IBM Cloud Academy. We are delighted to have you as a member of this community, and I look forward to working with you as you join with other leading academic institutions globally to shape the direction and implementation of Cloud Computing in Education.

As a member of the IBM Cloud Academy, you will have the opportunity to collaborate with others who are also integrating Cloud Computing into their campus infrastructures to share best *practices, lessons learned, innovative ideas and insights. You will be participating in the definition of emerging cloud technologies and implementations with the other members of the academy, as well as IBMers from research, development, and brands.* The initial K-12 and Higher Education members of the IBM Cloud Academy have articulated the benefits they see from membership in the program charter as follows:

- *Community Access.* Access to a support community that is changing local IT culture by embracing cloud computing.
- *Knowledge Sharing.* Thought leaders and technical experts with knowledge of proven cloud computing technologies and solutions.
- *Project Support.* Technical and business case examples, benchmarking data and access to resources that support cloud computing projects.
- *Project Funding.* Funding opportunities for cloud computing projects and research that are competitively awarded
- *Resources.* Strategies to use cloud computing to transform education and research in K-12 and higher education globally.
- *Headlights:* Trends in K-12 and higher education globally as relates to the move to cloud computing models and approaches.

You will be receiving further information on registration and training for participation in the 'collaboratory', used for meetings, discussions and projects. Based on the LotusLive solution available on the IBM Cloud, the collaboratory makes it easier for academy members to collaborate as a community virtually.

Welcome to the IBM Cloud Academy.

Sincerely,

Chris Bernbrock

Christopher W. Bernbrock
Program Director, IBM Cloud Academy
IBM Global Education

Fig. 5.3 Invitation to Cloud Academy

Fig. 5.4 IBM Cloud Academy

Fig. 5.5 IBM Opens the First Multipurpose Cloud Computing Center in Poland

Wrocław University of Technology and IBM Poland plan to continue collaboration in the field of education, PHD programs and research. IBM plans to initiate similar programs with other institutes, faculties and data centers in Poland, as well as to integrate with local business communities. World of Innovation - this is the title of the conference, concerning the latest achievements in the IT field, which took

Wrocław University of Technology

Rector

Wroclaw, 18 April 2012r.

Within a Committee on Cooperation with IBM Cloud Academy which was established on the grounds of the Circular Letter 51/2010, I appoint Mr. Jerzy Kotowski, PhD to represent Wrocław University of Technology in contacts with IBM Israel Science and Technology in Haifa in order to fulfill objectives defined in the Letter of Intent which was signed on April 3, 2012 and it was registered as the number P/55/12 in the Central Agreement Register

R E K T O R

Prof. Tadeusz Więckowski

Fig. 5.6 An appointment related to co-operation with IBM HRL

place 3rd April, 2012, in the Wrocław University of Technology (PWr) and was organized in cooperation with IBM Global Services Delivery Centre in Wrocław.

During the conference, Wrocław University of Technology and IBM Haifa Research Laboratory signed a Document of Understanding, concerning future co-operation. Israeli Center was represented by his director Oded Cohn, and the University by its Rector, Professor Tadeusz Więckowski. Collaboration will concern creating several laboratory groups for research and development of the latest IT technologies, in the field of Cloud Computing, IT modeling, as well as optimization problems. See Fig. 5.6.

One may say that the future of co-operation between Wrocław University of Technology and IBM Consortium is wide open. And we may easy assume that Cloud Computing will play an outstanding role as the useful, flexible end effective tool to support this co-operation.

Cloud computing can provide organizations with the means and methods needed to ensure financial stability and high quality service. Of course, there must be global cooperation if the cloud computing process is to attain optimal security and general operational standards. With the advent of cloud computing it is imperative for us all to be ready for the revolution. We will carry about this remark.

References

1. Averitt, S., Rindos, A., Vouk, M., et al.: Using VCL Technology to Implement Distributed Reconfigurable Data Centers and Computational Services for Educational Institutions. IBM Journal of Research and Development 53(4), 1–18 (2009)
2. Cagan, J., Shimada, K., Yin, S.: A survey of computational approaches to three-dimensional layout problems. Computer Aided Design 34(8), 597–611 (2002)
3. Dreher, P., Vouk, M.: Utilizing Open Source Cloud Computing Environments to Provide Cost Effective Support for University Education and Research. In: Chao, L. (ed.) Cloud Computing for Teaching and Learning: Strategies for Design and Implementation, pp. 32–49. IGI Global (2012)
4. Foy, N.: The Sun Never Sets on IBM. Morrow, New York (1975)
5. Garg, S.K., et al.: Environment-conscious scheduling of HPC applications on distributed Cloud-oriented data centers. J. Parallel Distrib. Comput. (2010), doi:10.1016/j.jpdc.2010.04.004
6. Greblicki, J., Kotowski, J.: Automated design of totally self-checking sequential circuits. In: Moreno-Díaz, R., Pichler, F., Quesada-Arencibia, A. (eds.) EUROCAST 2009. LNCS, vol. 5717, pp. 98–105. Springer, Heidelberg (2009)
7. Greblicki, J., Kotowski, J.: Analysis of the properties of the Harmony Search Algorithm carried out on the one dimensional binary knapsack problem. In: Moreno-Díaz, R., Pichler, F., Quesada-Arencibia, A. (eds.) EUROCAST 2009. LNCS, vol. 5717, pp. 697–704. Springer, Heidelberg (2009)
8. Kotowski, J.: The use of the method of illusion to optimizing the simple cutting stock problem. In: Proc. MMAR 2001, 7th IEEE Conference on Methods and Models in Automation and Robotics, vol. 1, pp. 149–154 (2001)
9. Rodgers, W.: Think: A Biography of the Watsons and IBM. Stein and Day, New York (1969)
10. Vouk, M.: Cloud Computing - Issues, Research and Implementations. Journal of Computing and Information Technology 16(4), 235–246 (2008)
11. Watson, T.J.: A Business and Its Beliefs. McGraw-Hill, New York (1963)
12. http://www.accountingweb.co.uk/topic/technology/one-hundred-years-business-machinery/506493
13. http://www.ibm.com/cloud-computing/us/en/?cm_re=masthead-_-solutions-_-cloud
14. IBM; Let's build a smarter planet, http://www.ibm.com/smarterplanet/us/en/?ca=v_smarterplanet
15. IBM Research; Explore our latest breakthroughs and innovations, http://www.research.ibm.com/featured/whats-new.shtml

Chapter 6
Development of Intelligent eHealth Systems in the Future Internet Architecture

Paweł Świątek, Krzysztof Brzostowski, Jarosław Drapała,
Krzysztof Juszczyszyn, and Adam Grzech

Abstract. One of the most important directions in the development of eHealth systems is their personalization. Body area networks with wireless sensors and portable access devices allow one to design applications for supporting people in their daily activities. On the other hand, they require substantial data analysis and processing which are inherent features of the medical domain. We describe few service-based applications utilizing a Future Internet architecture which supports QoS guarantees for the communication between their components and also show how they can benefit from its features such as quality of service (QoS) provisioning and resources reservation. We demonstrate how proposed applications address content, context and user awareness based on the underlying Next Generation Network (NGN) infrastructure and give examples of decision-making tasks performed by applications.

6.1 Introduction

Composition of Body Area Networks systems with the use of wireless sensors and smartphones makes it possible to:

- perform vital sign measurements *"anywhere and anytime"*;
- process measurement data on the smartphone or in a computer centre that provides services via a network.

These possibilities enable one to design intelligent eHealth applications. Roughly speaking, we understand the intelligence of the eHealth application as:

- distributed measurement data acquisition and processing;
- decision-making support;

Paweł Świątek · Krzysztof Brzostowski · Jarosław Drapała ·
Krzysztof Juszczyszyn · Adam Grzech
Institute of Computer Science, Wrocław University of Technology,
Wybrzeże Wyspiańskiego 27, 50-370 Wrocław, Poland
e-mail: {pawel.swiatek,piotr.klukowski,krzysztof.brzostowski,
        jaroslaw.drapala}@pwr.wroc.pl

© Springer International Publishing Switzerland 2015   73
R. Klempous and J. Nikodem (eds.), *Innovative Technologies in Management and Science*,
Topics in Intelligent Engineering and Informatics 10, DOI: 10.1007/978-3-319-12652-4_6

• ability to compose services tailored to the needs of users and the profile (context, content and user awareness)./indexuser awareness

All of the above require integration of service composition approaches with the communication QoS guarantees provided by the NGN infrastructure. For that reason, the eHealth domain was chosen to demonstrate the applicability of our approach.

Recently, both commercial and research teams develop solutions for supporting the everyday life of individuals (for professional and recreational purposes). Such systems provide wireless measurements of vital signs and tools to process them in order to support training and monitoring of one's state of health. In the commercial fields of applications Polar[1] belongs to the most important companies offering wearable devices and data processing tools. Wristwatches and chest-worn heart rate sensors are used to measure and present data to the user. Polar provides equipment for fitness improvement and for performance maximization. These devices are used in motivational feedback (i.e. generate beeps every time a certain amount of calories is burnt). It is useful in preventing injuries and overtraining. Moreover, Polar's software provides tools to optimize training intensity. Suunto[2] provides devices that generate a personalised training plan. These devices are capable of making recommendations for training volume (i.e. the frequency, duration and intensity). Additionally, a proposed training plan may be adjusted to the user's current capabilities (i.e. when user's activity level decreases).

The *miCoach*[3] product from Adidas is an advanced training tool that can be used to optimize one's training plan for endurance training, strength and flexibility. Data are measured based on an individual's stride and heart rate monitor. The website allows the user to manage the training process. An important feature of the system is *digital coaching* which serves to motivate the user via feedback by giving voice notifications to him/her such as "*speed up*" or "*slow down*" and giving information about the progress of the workout. MOPET [6] is an example of an advanced project under development. It uses measurement data to work out the user's mathematical model. The model adapts to the user and is used to predict the user's performance. On the basis of the model analysis, advice concerning the user's state of health and safety issues are generated. Such a model-based prediction is an important element of personalization and context awareness systems.

The crucial observation is that the functionalities of the applications described above are part of a particular system and cannot be directly used by other systems. At the same time, they do not assume an explicit usage of any given network infrastructure; hence they cannot benefit from its QoS provisioning and management. The main idea behind our proposition is the development of eHealth systems in such a way that they could share their functionalities in the form of services in the network,

---

[1] Polar, www.polar.fi, accesed: 4 April 2012.

[2] Suunto, www.suunto.com, accesed: 4 April 2012.

[3] Adidas miCoach, The Interactive Personal Coaching and Training System, www.micoach.com, accesed: 4 April 2012.

taking advantage of the *Service Oriented Architecture* (SOA) and the Future Internet architecture and profiting from the synergy between them.

One of the main motivations for designing new architectures for the Future Internet is to meet the challenges imposed on the ICT (Information & Communications Technology) infrastructure by new eHealth applications. These challenges include, among others:

1. *Content awareness* meaning the sensitivity of data processing and transmission methods to the content being delivered to the end-user. Content awareness may emerge in: different processing of various data streams (i.e. video encoding or sensor data encryption) and different forwarding methods (e.g. routing) for various streams.
2. *Context awareness* consisting of the different treatment (in terms of forwarding and processing methods) of traffic depending on the particular use-case scenario of the application generating this traffic. The context may be determined by a networking device or geographical localization of the user.
3. *User awareness* understood as personalization of services delivered to the end-user. Personalization is achieved by means of the proper choice of data processing and transmission methods according to functional and non-functional requirements stated by the user. The user's requirements may be formulated explicitly or be a result of an automatic recommendation which is based on the history of the application usage.
4. *Sensor networks*: vehicle networks, intelligent building infrastructure, telemedicine, etc. Each particular telemetry application involves specific types of data processing methods and the transmission of large number of small portions of data often requiring real-time or near real-time end-to-end performance.

Augmentation of the current Internet architecture with the abovementioned functionalities will fulfill the assumptions of the *pervasive computing* paradigm where end-to-end services delivery is facilitated by a cloud of distributed networking devices and loosely coupled application modules (domain-dependent services). The key feature of such an approach is the user centricity where the user does not invoke any particular application or service nor even specifies where the application should be executed.

Each application presented in this work is service based and benefits from the underlying IPv6QoS architecture which allows for QoS maintenance and guarantees. This feature is especially important when distributed, component-based applications are QoS-dependent and the simultaneous use of portable devices and complex data processing issues are involved. In this case, reliable communication services and appropriate computational resource allocation must be provided. The applications presented in this work are examples of such solutions. Their originality stems from the joint use of the NGN communication infrastructure and the service selection and composition techniques.

In Section 6.2 we give a brief overview of the architecture of the IPv6 QoS system with special emphais set on the service stratum signaling system in Subsection 6.2.2, which forms an NGN communication framework for distributed service-based applications. Next, in Section 6.3 we present an exemplary application and also show how content, context and user awareness are achieved by using service stratum signaling. Additionally, we show in Section 6.4 how a custom application can be designed using already existing processing and communication services in the system which effectively demonstrates how service composition schemes and the IPv6QoS NGN approach may be integrated. In Section 6.5, we conclude our work and point out directions for future work.

6.2 Systems Architecture

Currently there are a number of approaches aiming at meeting the assumptions of the Future Internet. Depending on the proposed ICT architecture some or all of them are assured by the utilization of the proper networking techniques and concepts. These approaches differ from others in the layer in which the new functionalities of the Future Internet are accomplished. As an example, consider the Content Centric Network (CCN) proposed by Van Jacobson [11] where content delivery mechanisms (e.g.: caching, forwarding, security, etc.) are mostly implemented at the lower network layers. This revolutionary post-IP approach requires the entire networking protocol stack to be redesigned. On the other hand, in the prototype of the Parallel Internet CAN (Content-Aware Network) [3], being one of the results of the Polish national project IIP (Polish acronym for Future Internet Engineering) [5], the content is encapsulated in the new frames format, but signaling messages are passed by using the IPv6 (Internet Protocol version 6) protocol, and content delivery route calculations and caching are accomplished at the application layer. These two approaches are candidates for implementing the concepts of the Internet of content and media. One of the most mature architectures for the Future Internet is the Next Generation Network (NGN) [10]. The NGN signaling system, in conjunction with the Differentiated Services (DiffServ) quality of service assurance model [4] and the IPv6 networking protocol stack, allow the implementation of a converged all-IP multi-domain network conforming with all assumptions of the Future Internet. A sample implementation of this approach is the Parallel Internet IPv6 QoS prototype [5, 21].

The main contribution of this work is the concept of how to achieve content, context and user awareness in the IPv6 QoS architecture by proper signalization in the service and transport stratum of the NGN architecture. The proposed concept is illustrated on exemplary applications designed for the IPv6 QoS system.

6.2.1 IPv6 QoS System

In this work we consider an IPv6 QoS system architecture developed in the Polish national project IIP [3]. In this architecture it is assumed that the system consists of multiple layers each of which provides certain functionalities to the adjacent upper layer. The first layer is a physical network infrastructure which with use of virtualization techniques [13] provides to the second layer a virtualized networking environment with dedicated communication and processing resources. Such virtualization allows for the coexistence of multiple isolated virtual networks (called parallel Internets – PI), characterized, among others, by different frame formats, protocol stacks and forwarding capabilities in a single physical infrastructure.

The IPv6 QoS system is one of the parallel Internets existing in a virtual networking environment. In general, the IPv6 QoS architecture is based on the coupling of the DiffServ quality of service assurance model and the NGN signaling system. DiffServ is responsible for delivery to traffic flows generated by users required level of the quality of services by means of flow admission control, classification of flows to pre-defined traffic classes and processing of aggregated flows from different traffic classes. The NGN signaling system is used to provide end-to-end QoS guarantees by reserving the necessary amount of communication resources to each particular connection request. The reservation of communication resources is performed by assignment of the request to the proper DiffServ traffic class which meets the QoS requirements for this flow.

The purpose of signaling in NGN is twofold. The first is to reserve the required communication resources and to establish an end-to-end connection between a pair of hosts in the system. This signaling is performed at the network layer in the so-called *transport stratum*. The second type of signaling is performed at the application layer (*service stratum*). Service stratum signaling is, in general, an application-specific signaling (e.g. SIP signaling), the aim of which is to configure distributed modules of an application and to process information necessary to send to transport stratum a request for communication resources reservation. Signaling can be also viewed as a middleware which separates the networking layer functionalities and application domain-specific specific functionalities.

6.2.2 Service Stratum Signaling

The task of service stratum signaling is to control the execution of distributed applications and to pass communication resources reservation requests from applications to the network. The service stratum, being an intermediate layer between applications and the network which translates application-specific signaling and negotiations to a uniform service stratum – transport stratum interface, allows the implementation of arbitrary application-specific signaling schemes. This in turn allows for the achievement of content, context and user awareness by the implementation of specialized services management mechanisms whose task is to transparently compose and control the execution of personalized complex services based on functional and non-functional requirements.

In our approach, based on the Service Oriented Architecture (SOA) paradigm, we assume that applications in the IPv6 QoS system consists of distributed loosely coupled modules (called atomic services). Execution of each application use-case scenario is performed by sending a request to an application server which composes a specialized complex service from the set of atomic services available in the system. Additionally, we assume that end-to-end communication services provided by the IPv6 QoS system are also treated as atomic services and can be utilized to deliver the requested complex services according to SOA approach to the user.

In order to deliver the requested complex services to the user, a two-stage signaling in the service stratum is proposed. The task of the first stage of signaling is twofold. Firstly, based on services available in the distributed communication system, it allows one to compose a complex service which conforms with the functional and nonfunctional requirements [9]. Secondly, it notifies each module of distributed application how and with which module they should establish communication with in order to deliver the requested complex service. The aim of second stage signaling is to negotiate the details of communication between each pair of atomic services taking part in the complex service delivery. Communication details depend on the functionalities of communicating services and may concern, among others: data formats, audio and video codec, required transport protocol, encryption, etc. Taking into account the negotiated communication details and non-functional requirements concerning the requested complex service, proper end-to-end communication service is requested from the IPv6 QoS system for each pair of communicating atomic services. Note that thanks to the negotiation process requested end-to-end communication services depend on the context of communication and the content being transmitted resulting in fully personalized context and content-aware complex services delivery.

An exchange of signaling messages is required to prepare exemplary service and consist of sending the requested data from a service to a user is presented in Fig. 6.1. During the first stage, an arriving service request (1) is processed by a server (2) in order to compose a complex service that conforms to the user's requirements.

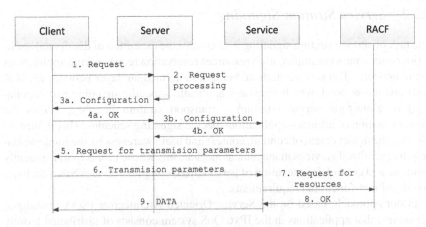

Fig. 6.1 Signaling messages exchange for an exemplary complex service

The result of a composition process is a set of atomic services available in a distributed system which will take part in the execution of a request (1). During the second stage, the server configures all the necessary atomic services (including a client application) by sending them information about: source of data, data processing methods and the destination of processed data (3a and 3b). In other words, during this stage each service is notified from who and what it will receive, what to do with received data and where to send it. After configuration is confirmed to the server (4a and 4b), each pair of communicating services negotiates values of application and communication-specific parameters (5 and 6). When communicating parties agree upon the type of communication-appropriate end-to-end communication services (which guarantee the delivery of the required QoS parameters for data transfer), it is requested from the IPv6 QoS system. This is accomplished by sending a resources reservation request to the resources and admission control function (RACF) (7) with use of a service control function (SCF). After confirmation of connection preparation (8), data transmission may take place (9). The first stage of signaling beginning with the sending of a request (1) and ending with the configuration of all services (4a and 4b) is accomplished with the use of an XML-RPC protocol. The second stage of signaling, consisting of the negotiation of communication values and other application-specific parameters (5 and 6), is accomplished with the use of XMPP protocol. It is important to note that vertical communication (signaling) of application components and the network is done with the use of service stratum transport stratum interface (SCF RACF to be exact). This means that each application component that is able to send requests to the network should be equipped with an SCF module which translates application-specific horizontal signaling to application-independent uniform vertical signaling between the service and transport stratum defined by the SCFRACF interface.

6.3   Future Internet Applications

Three applications are described here. The eDiab and SmartFit are eHealth applications that provide advanced functionalities for sportsmen and patients, and an Online Lab that may provide computational services for another applications.

6.3.1   eDiab

The eDiab application is designed to support the everyday life of diabetic patients (Type 2 diabetes). The user gains access to:

- **communication services** that deliver measurement data and results of their processing to the destination;
- **computational services** that perform advanced data processing.

Data Acquisition is performed by wireless sensors (glucometer, pressure-gauge). Sensors send data via Bluetooth directly to the smartphone. In the eDiab application, we use the GlucoTel glucometer and PressureTel pressure-gauge, connected to the smartphone with Symbian OS. Next, request for communication services or computational services is generated and sent from the smartphone to the eDiab server. The request is processed by the Universal Communication Platform [19], performing:

- service stratum signaling;
- negotiation of connection parameters with the eDiab application server;
- service composition.

After the connection is established, the UCP mediates data transfer between services. Data are sent to the appropriate services, stored in a database and made accessible for authorized users in the web service. Other services process the data and work out feedback response to the user. The network topology of eDiab is given in Fig. 6.2. In order to communicate with the Future Internet network, the eDiab server negotiates and transfers data with the use of XMPP protocol. The following stages of data processing take place in the system:

- measurement data acquisition;
- classification of incoming data (*hypoglycemia*, *low*, *normal*, *high*, *hyperglycemia*);
- estimation of the patients model parameters;
- advice concerning current physical activity, diet and control of blood glucose level (BGL).

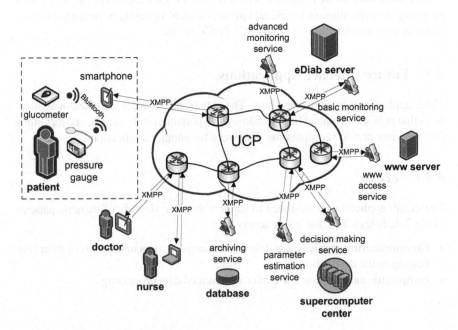

Fig. 6.2 Network topology for eDiab

Measurement data are classified by the rule based classifier (rules are defined in the Drools Business Logic Integration Platform). Classification results are stored in a database and are available through the web service. If measurements fall into the hypoglycemia class, an alert is generated. If the class is different than normal, an email notification is sent to a nurse or a doctor.

Advice, decisions and notifications worked out by computational services are delivered to the smartphone. They are in the form of voice commands, e.g. ,"eat something", ,"slow down", or ,"you may keep the current exercise intensity for about 5 minutes". All decisions are generated or validated on the basis of the Glucose-Insulin model simulations [8]. The simulations may be performed by computational services provided by OnLine Lab. The model is a part of the user profile and contains parameters that should be estimated from the measurement data. In order to fit the model to the user, appropriate computational services are employed. Then, the user profile is updated (personalized).

Diabetic patients are recommended to practice regular physical exercise in order to lose weight. Since during physical activity the glucose utilization rate is high, it is important to control the intensity of exercise [12]. The goal of control is to keep the blood glucose level within the normal range during exercise. The proposed system architecture allows one to easily integrate the SmartFit application to support training with the eDiab application to control one's blood glucose level. To make use of their services is straightforward in order to provide a new complex service to a diabetic user who wants to perform physical exercise. Services and their parameters may be set in accordance with the current state of health of the patient (*user awareness*). The task that the patient performs is associated with a use-case scenario which further imposes the choice of services and their parameters (*context awareness*). Moreover, some data, such as personal health records and blood glucose level measurements, have to be encrypted when transferred through the network (*content awareness*) [20].

All the mentioned eDiab functionalities may be delivered to the user thanks to the service oriented system architecture, which allows one to compose different complex services on the basis of atomic services [9]. Composition is about network and computational resources reservation and preparation. The systems may adapt to the patient and the activities he/she performs [18, 22, 23, 24, 25, 26].

The eDiab utilizes the GlucoTel[4] blood glucose meter with Bluetooth technology. The BodyTel also provides a mobile app for Symbian and Android devices and an online portal with patient records. However, it mainly serves to record data. It does not perform advanced processing and decision support. A similar application is OneTouch from the LifeScan company[5]. A more advanced solution is provided by the MyGlucoHealth[6], where reminders, messages and alerts may be generated.

---

[4] https://secure.bodytel.com/en/products/glucotel.html, accesed: 20 April 2013.

[5] http://www.lifescan.pl/, accessed: 20 April 2013.

[6] http://www.myglucohealth.net/index.html, accesed: 20 April 2013.

One of the most comprehensive and advanced products are offered by the WellDoc[7] and Sanofi BGStar[8]. The WellDoc includes automated coaching that uses system analysis, data mining and decision support techniques. Both producers cooperate and exchange their knowledge and experience[9]. The eDiab is designed to be adaptive, meaning that it keeps the quality of services at the desired level. Unlike the applications described above, it is able to reconfigure in response to variations in network traffic [29, 30].

6.3.2 SmartFit

Developing wireless sensor networks, wearable sensor technology and methods of fusion with information techniques provides the basis for pervasive computing systems. In such systems, wireless sensor networks are connected with a number of low-cost, low-power sensing nodes and, in addition, they can use various computational services for data transfer, processing, storage and decision-making support [2]. Pervasive computing systems, which can operate in distributed environment, are powerful tools to be used in various areas of application e.g. healthcare, sports, industry or entertainment.

SmartFit is a system that incorporates new information and communication technologies for pervasive computing. The main functionalities of the system concern the physical and technical training of athletes. One of the main non-functional features of application is the ability to provide its functionalities *"anywhere and anytime"*. This means that acquired data must be transmitted between all users of the system (i.e. athletes and coaches), no matter their location. Moreover, it is crucial to transmit data with a predefined level of quality. To this end each functionality was decomposed to small modules called atomic services. Each atomic service has few different versions with predefined levels of quality. These different versions of atomic services are used in the process of user-centric functionality delivery with the use of an orchestration mechanism. User-centric functionality means that in order to compose the required functionality, the user's specific requirements and needs are taken into account.

In Fig. 6.3 the general architecture of SmartFit is presented. It is a classical three-tiered architecture. The first tier is used to acquire sensor data. The second tier is designed for data processing and decision making. The last one is used to present results of data processing and decision making to the user. For each tier the set of atomic services is defined. In the process of user-centric functionality composition, all versions of atomic services at each tier are taken into account. One of the main functionalities of the system concerns planning the volume of physical training taking into account long-term performance effects. This functionality can be used by

---

[7] http://www.welldoc.com/Products-and-Services/
Our-Products.aspx, accesed: 20 April 2013.

[8] http://www.bgstar.com/web/, accessed: 20 April 2013.

[9] http://mobihealthnews.com/17189/
apple-stores-now-sell-sanofis-iphone-glucose-meter/

athletes in various sports. Another functionality of the system is related to physical training monitoring. It is used to supervise correctness of the performed exercises, their order and number of repetitions. Moreover, it can be used to predict the occurrence of injuries.

Another functionality of the SmartFit system is to support technical training. The architecture of the SmartFit system provides a mechanism for adding new functionalities easily. To this end it is necessary to prepare new atomic services that will be used to compose complex service to support technical training for a particular discipline. We focused on technical training in tennis.

In Fig. 6.4 the network topology is presented. The main element of this network is the SmartFit server that supervises the process of problem-specific functionality delivery. This process has three phases. During the first one, the server must configure all the necessary atomic services. In the considered example, the role of these atomic services is to:

- facilitate physical and kinematic data acquisition from the user's Body Area Network (BAN);
- detection of changes;
- extraction of features from acquired data;
- pattern classification.

The second phase is signalization with the use of SIP and XMPP protocols between services distributed in the network. The last phase of the functionality delivery process is data transmission, processing and presentation (to the user, trainer, doctor, etc.). A skill assessment task allows one to determine the current skill level of a tennis player. Results are helpful for making recommendations about the future technical training of tennis stroke (e.g. serve, forehand and backhand) planning. Tennis player-oriented technical training is conducive to skill acquisition and improvement.

The tennis player should be equipped with wireless gyroscopes. Physiological and kinematic data are transferred to the server and processed by the classification

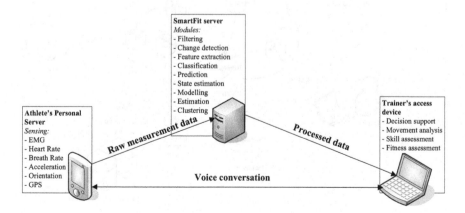

Fig. 6.3 Architecture of SmartFit system for distributed environment

service. The tennis player is assigned to a predefined skill level group. It is also possible to identify elements of a tennis stroke that are performed incorrectly. Based on measurement data, a personalised and mathematical model of the player is worked out. On the basis of the model recommendations for future technical training are suggested.

Another functionality of SmartFit is supporting the tennis player in feedback training. Feedback training helps one to master tennis technique and skills. Physiological and kinematic data from sensors placed on the athlete's body are monitored during movements in a real-time environment. SmartFit supports both the tennis player and the trainer in comparing the player's performance to reference data acquired from previous training or from elite tennis players.

The system may also utilize EMG measurement data, if available. An EMG signal is transferred to the server and filtered by an appropriate computational service. The results are sent to the trainer's portable access device, and the muscle activation sequence for upper arm during strokes is presented. Signals from gyroscopes and accelerometers are used to estimate the trajectory of upper arm movement during tennis strokes such as during the serve, forehand and backhand. Moreover, parameters of the upper arm movement's mathematical model are determined for personalization purposes.

In order to provide the required quality of service it is necessary to apply a context awareness mechanism. Context awareness incorporated in SmartFit is used to

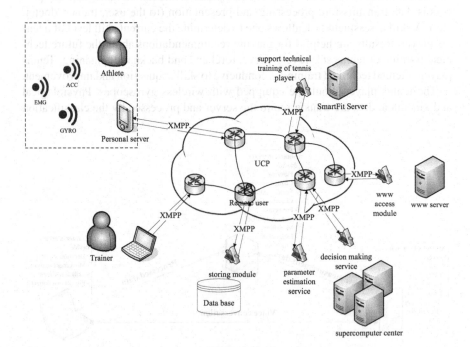

Fig. 6.4 Network topology of the system for technical training support

adapt the packet size according to the user's requirements. Context information can be obtained through sensor networks, e.g. measurement of heart rate during training session or/and personal server e.g. GPS or networking devices. Based on this information it is possible to predict user behaviour and his/her location.

6.3.3   Online Lab

Functionality

The virtual laboratory infrastructure should automate most tasks related to the reliable and reproducible execution of required computations [14], which premises were implemented in the Online Lab architecture. It is a distributed, extensible, service-based computational laboratory benefiting from the IPv6 QoS architecture used to distribute computational tasks while maintaining the quality of service and user experience [15]. It allows its users (i.e. students or researchers) to access different kinds of mathematical software via Python math libraries and perform computations on the provided hardware without the need for installing and configuring any software on a local computer. The communication mechanisms are designed for optimization of the user's Quality of Experience, measured by the response delay. The functionality of the Online Lab embraces:

- access to computational services ensured by the user's virtual desktop which is a windowed interface opened in a Web browser and implemented in JavaScript;
- creation and removal of computational services with no limitations being assumed on the nature of computations – the users may freely program computational tasks in any language interpreted by running computational services (in the first prototype a set of Python engines equipped with the math libraries was provided);
- user profile maintenance and analysis – the users are distinguished by their profiles which hold information about their typical tasks and resource consumption.

Online Lab (OL) implements an architecture consisting of a user interface (OL-UI), core server (OL-CORE), services and computational engines. OL-UI is a web service emulating a desktop and a window manager. Code is being typed into specialized data spaces - notebooks, which are executable documents executed by OL-Services.

The process of a user's query execution is presented in Fig. 6.5. OL-Core and OL-Services belong to the service. One notebook represents one computational task. The system also may recommend notebooks of other users. The content of the notebooks is annotated with the help of domain (Math) ontology. The OL-Core is constantly monitoring the OL-Services, storing execution times and data transfers in its database. The memory and CPU usage of the engine processes are also being monitored and stored. From the user point of view, in the case of computational tools, the key element of the Quality of Experience (QoE) is waiting time which is the sum of computation time and communication times. The first is query-specific and must be taken into account as a value predicted on the basis of known history of user queries. The second depends on the volume of data and code and influences the link parameters negotiation performed by the Online Lab Core.

Awareness Capabilities of the OnlineLab

The Online Lab classifies user queries (computational tasks) and reserves the communication services of the IIP system in order to guarantee the QoE for the user. The computational tasks are scheduled in order to minimize the wait time of the user which is done by computational service monitoring and dynamic configuration of communication links using the IPv6 QoS infrastructure. This approach is used to address the requirements defined in the introductory section:

1. *Content awareness* - such an OL-Service is chosen to provide the minimum processing time. The data volume of the task influences the parameters used during the link reservation in the IPv6 QoS system (to achieve the minimum transfer time).
2. *Context awareness* is maintained by the Load Balancer. Its task is to analyze the stream of requests and manage the negotiations with the service stratum. It is also equipped with the prediction module which forecasts user behavior which allows for the prediction of the computational and communication resources as well.
3. *User awareness* – the services are personalized, taking into account the user's preferences, typical volumes of associated data and a recommendation scheme.

Taking the above into account, the general task of the Online Lab is to compose a computational service, given the request stream from the users is known or predicted. All the components of the system (OL-Core and available OL-Services) are registered and have unique IDs. Once the optimal (with respect to the QoE) structure of this complex service (including the set of OL-Services and the parameters of communication links between them) is decided by the Load Balancer, the OL-Core

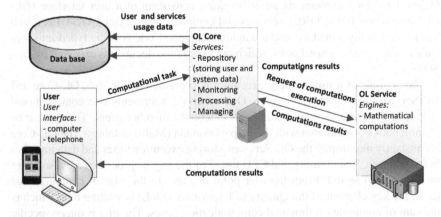

Fig. 6.5 The general schema of the Online Lab service execution

reserves (via SCF functions, as described in Sect. 6.2) the communication links connecting all the Online Lab services. This guarantees delivery of the required QoS parameters. In the second phase, the negotiation in the service stratum takes part in establishing and confirming pairwise communication between the services. After that, the computational tasks are scheduled and assigned to the appropriate OL-Services by the OL-Core.

An additional unique feature of the Online Lab is the possibility of implementing dedicated computational services which may be available for other applications. An example of this scenario will be sketched in the following section where we describe the use of the Online Lab service to be used by the SmartFit application.

6.4   Custom Application

The complex service model assumed for IPv6 QoS applications makes it easy to develop intelligent applications built on the basis of available communication and computational services. Below, we describe the custom application that serves to **support endurance training for a runner**.

6.4.1   Problem Formulation

The goal of the footrace is to run a given distance in the shortest time. Two variables are measured by wireless sensors in real-time: the user speed and heart rate (HR). The most important issue in this task is to spread the effort of a runner in such a way that the time of the run is the shortest. The system does it for the user. All decisions are worked out on the basis of a mathematical model describing the cardiovascular system and fatigue during exercise [7]:

$$x_1'(t) = -a_1 x_1(t) + a_2 x_2(t) + a_2 u^2(t), \tag{6.1}$$

$$x_2'(t) = -a_3 x_2(t) + \frac{a_4 x_1(t)}{1 + \exp(a_5 - x_1(t))}, \tag{6.2}$$

where x_1 is the heart rate change from the rest (resting heart rate), u denotes speed of the exerciser, x_2 may be considered as fatigue and parameters $a_1, a_2, \ldots, a_5$ take nonnegative values. Each user performance may be characterized by the set of 5 numbers. The model parameters are stored in the user profile.

Typical training protocol involves step-like functions $u(t)$ that determine length of the resting (zero speed), the exercise (high speed) and the recovery (walking or resting) periods (see Fig. 6.6).

6.4.2 Computational Services

The basic computational service is a differential equations solver (*DES*). It's role is to work out the model response (x_1 and x_2 time courses) for a given training protocol. After it receives the user parameters, initial conditions $x_1(0)$,$x_2(0)$ and training protocol $\{u(t)\}_{t=0}^{T}$, it solves the system of equations (6.1)-(6.2) using the Runge-Kutty method and sends $\{x_1(t)\}_{t=0}^{T}$, $\{x_2(t)\}_{t=0}^{T}$ back to the service that requested the solution.

The computational service that needs *DES* service to work accurately is the system identification service (*SIS*). It's role is to adjust the model to the user. The service receives measurement data, performs the least square approximation on demand and sends the resulting values of the model parameters to a service responsible for the user profile update.

The model validation service (*MVS*) returns the result of the binary classification: whether the parameters of the user model are up to date or not. The service is fed by the latest measurement data and current user profile. It uses *DES* service simulations and statistical inference methods to validate the model. If the user model responses become significantly different than the measured ones, *PES* should be run to update the parameters.

The training protocol optimizer (*TPO*) is a computational service that returns optimal protocol for a given footrace and the user. It solves the following optimization task. As an input data, it takes: distance D to run, the model of the exerciser, fatigue and speed limits $x_2^{\max}$ and $u^{\max}$ of the exerciser. It returns such a training protocol $u^*(t)$, for which a desired distance is completed in the shortest time T^*. The footrace time T is the solution to the equation:

$$D = \int_0^T u(t)dt. \qquad (6.3)$$

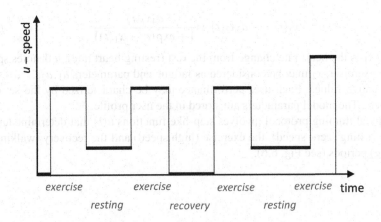

Fig. 6.6 Typical training protocol

There is a constraint for fatigue:

$$\max_{t\in[0,T]} x_2(t) \leq x_2^{\max},$$ (6.4)

[note that x_2 is related to $u(t)$ by the model equation (6.2)] and constraint for the user speed:

$$\max_{t\in[0,T]} u(t) \leq u^{\max}.$$ (6.5)

In order to solve the optimization task described above, *DES* must be available. It is employed to simulate the user's behavior for training protocols that are tried out during successive iterations of the optimization procedure. Each run of *TPO* requires multiple runs of *DES*.

Optimal training protocol $u^*(t)$ must be actuated. The user is expected to follow $u^*(t)$ as close as possible. He/she should be supported in:

- switching between exercise;
- resting and recovery periods;
- maintaining correct speed during these periods.

Typical control signals involve:

- voice commands uttered by the smartphone (e.g. *"run faster"*, *"slow down, please"*);
- music volume;
- frequency of the *"beep"* sound.

The choice of the control signal depend on the user preferences. Actuation of the training protocol is supported by the speed control service (*SCS*).

Since *DES* and *SCS* are simple procedures, their executable files may be sent to the smartphone and are then run. On the other hand, *MVS*, *SIS* and *TPO* need high computational power; therefore, they need to be run in a computer center. The smartphone only receives results of computations.

6.4.3   Decision-Making Support

The system supports the user in controlling the intensity of an exercise. In this custom application, the data processing needed to support real-time training procedures is performed by a dedicated OL-Service which serves as a decision-making component of the SmartFit application.

The user's story is the following:
Before the application is used to support physical activity, the user is asked to perform a few predefined training protocols. This is the first adjustment of the model for the user and SIS is executed. Typical usage scenario starts from typing:

- the footrace distance;
- maximum speed;
- the limit of fatigue (maximum speed and fatigue limit are learned by the system later on).

TPO works out an optimal training protocol on the basis of the user model simulations made by *DES*. The training protocol is actuated by *SCS*. In the background, *MVS* compares the model responses to the measured signals. If the difference becomes significant, *SIS* is called to update the user model and then *TPO* is run once again with the new user model. This may happen mainly due to: worse or better (compared to average) user condition (which takes place very often after the party) and problems with tracking the training protocol (e.g. after the fall). Each time the *MVS* detects that the user does not follow the reference trajectory, *TPO* recalculates the optimal training protocol for the remaining part of the route.

With time, the user improves (or decreases) his/her performance and the model updates are necessary. Therefore, from time to time, *MVS* checks the user model validity and, if necessary, *SIS* is called to update the user model parameters.

All the mentioned services provide an adaptive control system [17, 27, 28] supporting the user in a real-time (see Fig. 6.7) configuration. Adaptation takes place in response to events occurring during the single training (context awareness) as well as a reaction to the long term-effects caused by systematic training (user awareness). It must be noted that the proposed application is just an example of a personalization scheme which relies on the composition of new services on the basis of predefined services (*SIS, TPO, DES, SCS, MVS*). The final functionality of the configuration parameters of services are suited to the user requirements. Configuration of services depends on the sport discipline practiced by the user (context awareness). The system reacts to the long-term effects caused by systematic training by adjusting the user model and the choice of services he/she needs (user awareness). Moreover, it is possible to deliver more advanced applications, making use of services derived from other applications. For instance, when we want to deliver the same applications for a diabetic user, we may compose a new service that supports physical training, taking the blood glucose level into account (see Fig. 6.8). In such a case, additional services to keep the blood glucose level within the normal range is executed and there are additional requirements to blood glucose data encryption (*content awareness*).

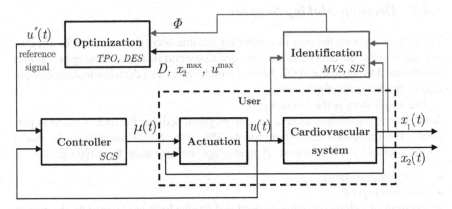

Fig. 6.7 The adaptive control system performed by computational services delivered by the OnLine Lab and executed within the SmartFit framework

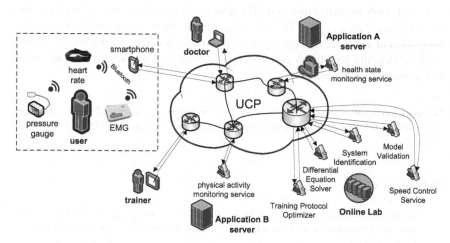

Fig. 6.8 Network topology for exemplary application

Additional constraint for *TPO* is defined to make sure that the training scenario generated by *TPO* will be safe for the user.

6.5   Summary and Future Works

In this work we presented a general idea of delivering to the end-user complex services in a distributed networking environment. The main feature of the proposed idea is that the process of complex services delivery is aware of the content being delivered, the context of the services delivery and that the delivered services are personalized for each separate end-user.

The proposed signaling system was designed and implemented as a middleware between the end-user and the network, more specifically as a service stratum in the NGN architecture of the IPv6 QoS system. Note, however, that the signaling system architecture does not assume any specific network architectures. The idea is to utilize communication services provided by the network layer to provide fully customizable application layer services built from computational and communication resources available in the distributed system. The benefits of the integration of NGN services and service composition and reusability schemes were illustrated by the examples from the eHealth domain which is characterized by strong communication QoS requirements and demands extensive data processing and analysis

With the proposed architecture it is easy to develop customizable eHealth applications. Depending on the set of services at hand, one can design an application suited best to current requirements. Moreover, within a single application the user is able to make use of all the available services, allowing one to combine different functionalities into one advanced component-based solution.

The IP QoS infrastructure and all applications presented in this work are the result of the IIP project [20]. The ongoing research efforts are devoted to the further integration of service-based applications within this scheme and the testing of new software-building paradigms and concepts stemming from service customization and composition approach.

Acknowledgements. The research presented in this paper was supported by the European Union within the European Regional Development Fund program no. POIG.01.01.02-00-045/09.

References

1. Ahmadi, A., Rowlands, D., James, D.A.: Towards a wearable device for skill assessment and skill acquisition of a tennis player during the first serve. Sports Technology 2, 129–136 (2009)
2. Alemdar, H., Ersoy, C.: Wireless Sensor Networks for healthcare: A Survey. Computer Networks 54, 2688–2710 (2010)
3. Bęben, A., Śliwiński, J., Krawiec, P., Pecka, P., Nowak, M., Belter, B., Gutkowski, J., Łopatowski, Ł., Białoń, P., Batalla, J.M., Drwal, M., Rygielski, P.: Architecture of content aware networks in the IIP system. Przegląd Telekomunikacyjny, Wiadomości Telekomunikacyjne 84, 955–963 (2011) (in Polish)
4. Blake, S., Black, D., Carlson, M., Davies, E., Wang, Z., Weiss, W.: An architecture for differentiated services, RFC2475 (1998),
   http://www.ietf.org/rfc/rfc2475.txt
5. Burakowski, W., Bęben, A., Tarasiuk, H., Śliwiński, J., Janowski, R., Batalla, J.M., Krawiec, P.: Provision of End-to-End QoS in Heterogeneous Multi-Domain Networks. Annals of Telecommunications 63, 559–577 (2008)
6. Buttussi, F., Chittaro, L.: MOPET: A Context-Aware and User-Adaptive Wearable System for Fitness Training. Artificial Intelligence In Medicine 42, 153–163 (2008)
7. Cheng, T.M., Savkin, A.V., Celler, B.G., Su, S.W., Wang, L.: Nonlinear Modeling and Control of Human Heart Rate Response During Exercise With Various Work Load Intensities. IEEE Trans. on Biomedical Engineering 55, 2499–2508 (2005)
8. Cobelli, C., Man, C.D., Sparacino, G., Magni, L., De Nicolao, G., Kovatchev, B.P.: Diabetes: Models, Signals, and Control. IEEE Reviews in Biomedical Engineering 2, 54–96 (2009)
9. Grzech, A., Rygielski, P., Świątek, P.: Translations of Service Level Agreement in Systems Based on Service-Oriented Architectures. Cybernetics and Systems 41, 610–627 (2010)
10. ITU-T Rec. Y. Functional requirements and architecture of next generation networks (2012),
    http://www.itu.int/itu-t/recommendations/rec.aspx?id=10710
11. Jacobson, V., Smetters, D.K., Thornton, J.D., Plass, M., Briggs, N., Braynard, R.: Networking Named Content. Communications of the ACM (2012), doi:10.1145/2063176.2063204

12. Man, C.D., Breton, M.D., Cobelli, C.: Physical activity into the meal glucose-insulin model of type 1 diabetes: in silico studies. Journal of Diabetes Science and Technology 3, 56–67 (2009)
13. Mosharaf, N.M., Chowdhury, K., Boutaba, R.: A survey of network virtualization. Computer Networks: The International Journal of Computer and Telecommunications Networking 54, 862–876 (2010)
14. Noguez, J., Sucar, L.E.: A Semi-open Learning Environment for Virtual Laboratories. In: Gelbukh, A., de Albornoz, Á., Terashima-Marín, H. (eds.) MICAI 2005. LNCS (LNAI), vol. 3789, pp. 1185–1194. Springer, Heidelberg (2005)
15. Pautasso, C., Bausch, W., Alonso, G.: Autonomic Computing for Virtual Laboratories. In: Kohlas, J., Meyer, B., Schiper, A. (eds.) Dependable Systems. LNCS, vol. 4028, pp. 211–230. Springer, Heidelberg (2006)
16. Porta, J.P., Acosta, D.J., Lehker, A.N., Miller, S.T., Tomaka, J., King, G.A.: Validating the Adidas miCoach for estimating pace, distance, and energy expenditure during outdoor over-ground exercise accelerometer. International Journal of Exercise Science 2 (2012)
17. Świątek, J.: Selected problems of static complex systems identification. Publishing House of Wroclaw University of Technology (2009) (in Polish)
18. Świątek, J., Brzostowski, K., Tomczak, J.M.: Computer aided physician interview for remote control system of diabetes therapy. In: Advances in Analysis and Decision-Making for Complex and Uncertain Systems, vol. 1, pp. 8–13 (2011)
19. Tomczak, J.M., Cieślińska, K., Pleszkun, M.: Development of Service Composition by Applying ICT Service Mapping. In: Kwiecień, A., Gaj, P., Stera, P. (eds.) CN 2012. CCIS, vol. 291, pp. 45–54. Springer, Heidelberg (2012)
20. Świątek, P., Juszczyszyn, K., Brzostowski, K., Drapała, J., Grzech, A.: Supporting Content, Context and User Awareness in Future Internet Applications. In: Álvarez, F., et al. (eds.) FIA 2012. LNCS, vol. 7281, pp. 154–165. Springer, Heidelberg (2012)
21. Tarasiuk, H., Śliwiński, J., Arabas, P., Jaskóła, P., Góralski, W.: Performance Evaluation of Signaling in the IP QoS System. Journal of Telecommunications and Information Technology 3, 12–20 (2011)
22. Tomczak, J.M., Gonczarek, A.: Decision rules extraction from data stream in the presence of changing context for diabetes treatment. Knowledge and Information Systems 34(3), 521–546 (2013)
23. Tomczak, J.M.: On-Line Change Detection for Resource Allocation in Service-Oriented Systems. In: Camarinha-Matos, L.M., Shahamatnia, E., Nunes, G. (eds.) DoCEIS 2012. IFIP AICT, vol. 372, pp. 51–58. Springer, Heidelberg (2012)
24. Rygielski, P., Tomczak, J.M.: Context change detection for resource allocation in service-oriented systems. In: König, A., Dengel, A., Hinkelmann, K., Kise, K., Howlett, R.J., Jain, L.C. (eds.) KES 2011, Part II. LNCS, vol. 6882, pp. 591–600. Springer, Heidelberg (2011)
25. Prusiewicz, A., Zięba, M.: Services recommendation in systems based on service oriented architecture by applying modified ROCK algorithm. In: Zavoral, F., Yaghob, J., Pichappan, P., El-Qawasmeh, E. (eds.) NDT 2010. CCIS, vol. 88, pp. 226–238. Springer, Heidelberg (2010)
26. Prusiewicz, A., Zięba, M.: The proposal of service oriented data mining system for solving real-life classification and regression problems. In: Camarinha-Matos, L.M. (ed.) Technological Innovation for Sustainability. IFIP AICT, vol. 349, pp. 83–90. Springer, Heidelberg (2011)

27. Brzostowski, K., Drapała, J., Grzech, A., Świątek, P.: Adaptive decision support system for automatic physical effort plan generation – data-driven approach. Cybernetics and Systems 44(2-3), 204–221 (2013)
28. Świątek, P., Stelmach, P., Prusiewicz, A., Juszczyszyn, K.: Service composition in knowledge-based SOA systems. New Generation Computing 30(2), 165–188 (2012)
29. Grzech, A., Świątek, P., Rygielski, P.: Dynamic resources allocation for delivery of personalized services. In: Cellary, W., Estevez, E. (eds.) Software Services for e-World. IFIP AICT, vol. 341, pp. 17–28. Springer, Heidelberg (2010)
30. Rudek, R., Rudek, A., Kozik, A.: The solution algorithms for the multiprocessor scheduling with workspan criterion. Expert Systems with Applications 40(8), 2799–2806 (2013)

Chapter 7
Understanding Non-functional Requirements for Precollege Engineering Technologies

Mario Riojas, Susan Lysecky, and Jerzy W. Rozenblit

Abstract. The design of accessible learning technologies for precollege engineering education is a multi-faceted problem that must take into account a multitude of physical, social, and environmental factors. Using literature reviews and assessment by a participant observer during an 18-hour intervention with a local middle school, we propose that the elicitation of non-functional requirements for precollege learning technologies can be better understood by dividing schools in clusters which share similar resources and constraints. Developers can utilize the proposed scheme as a means to establish minimal criteria that learning technologies must satisfice to be viable for adoption by a wider range of users and better meet the needs and priorities of students and educators.

7.1 Introduction

Precollege engineering education is a constructive discipline requiring both conceptual and procedural proficiency. The development of conceptual knowledge allows an individual to think about constructs in concrete and abstract ways. For example, considering the personal computer, abstract conceptual knowledge may relate to a system that processes and displays information, while concrete conceptual knowledge relates to the computers components, the function of the individual components, and the relationships among components. Moreover, a generalized understanding of systems independent of domain-specific knowledge (e.g. personal computer) demonstrates a deeper conceptual understanding. An analogous distinction exists between individuals with low and high levels of procedural knowledge. Individuals with a low level of procedural knowledge act randomly or based on

Mario Riojas · Susan Lysecky · Jerzy W. Rozenblit
The University of Arizona, USA
e-mail: mriojas@email.arizona.edu,
        {slysecky,jr}@ece.arizona.edu

© Springer International Publishing Switzerland 2015
R. Klempous and J. Nikodem (eds.), *Innovative Technologies in Management and Science*,
Topics in Intelligent Engineering and Informatics 10, DOI: 10.1007/978-3-319-12652-4_7

intuition, while high-level performers accomplish complex tasks with significant understanding of the separate decisions and steps taken to achieve their goals (Star 2000). With regard to engineering education, to produce effective solutions, learners must not only understand the underlying principles and theories (i.e., conceptual knowledge), but must also have the dexterity to nurture these solutions from a mental construct to a physical or virtual implementation procedural knowledge.

An elementary definition of engineering practice includes systematic approaches to problem-solving, specialized knowledge (e.g. design principles, systems theory, math- and science-based knowledge), and the ability to integrate and develop technology-based products and solutions (Sheppard et al. 2006). As a result, learning technologies in engineering education serve not only to assist curriculum and instruction in conveying conceptual understanding, they are also essential to developing procedural knowledge. The value of engineering education at the precollege level has been acknowledged by the National Academy of Engineering (Katehi et al. 2009) and recently by the National Research Council as part of its framework for precollege science standards (NRC 2011). However, it remains unclear how prepared the average school is for meeting the challenges of teaching engineering concepts and practices. While learning technologies abound providing alternatives ranging from open-source modeling software and open-hardware microcontroller kits to off-the-shelf robots and sophisticated computer-enhanced construction kits we argue that because many of these resources are not sensitive to quality requirements and constraints they are not as ubiquitous as they could be (Riojas et al. 2012). Especially at the middle-school level (grades 6th to 8th), where out-of-field teachers are often responsible for courses they lack background in (Peske and Haycock 2006), access to adequate curriculum and learning technologies is problematic. Moreover, dedicated engineering laboratories are uncommon at the middle-school level (Foster 2005).

Functional requirements refer to the functionality and behavior of a system, that is, actions a system is expected to perform given an input or stimuli. Performance requirements are described by the products time and space boundaries. Functional requirements and performance requirements typically serve as measures for technical excellence. In contrast, non-functional requirements capture quality characteristics such as usability, portability, or maintainability. The proper elicitation of requirements affects the quality of use as well as how successfully a product will satisfy user needs (Fig.7.1) (Glinz 2007). Bevan (1995) defines quality as the extent to which a product satisfies stated and implied needs when used under stated conditions. While technical functionality will always be a priority in learning technologies, it is equally important to emphasize quality-in-use and the satisfaction of user needs across a variety of work environments. Products that are considered examples of high technical excellence are not necessarily used at high rates. Adoption of quality-in-use measures within the domain of educational technologies will likely increase the satisfaction level of users (Bevan 1999).

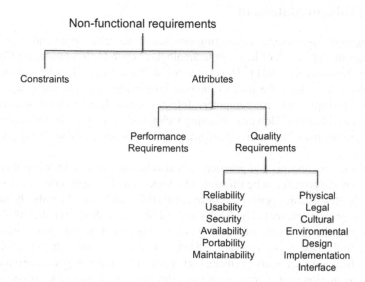

Fig. 7.1 Example of non-functional requirements

Incorporating non-functional requirements at an early stage of development would enable designers of learning technologies to achieve satisficing solutions for representative precollege school clusters, following the description in Lu (2009), of the balance between collective rationality and optimality in a complex system. In socio-technical problems, global optimality does not necessarily entail finding an unequivocal best choice for each stakeholder in the universe of possible solutions. Rather, approximations of a solution are provided to which heuristics can be applied to identify satisficing solutions. Because engineers are typically working with partial information and uncertainties in socio-technical problems, a way to achieve global optimality could be finding the best alternative from a set of solutions that satisfice a fundamental collection of requirements stated by stakeholders in the most rational manner.

A collective rationality that elucidates the needs and constraints that rule representative school clusters must be established to enable the development of technologies. Such technologies should have greater odds of meeting the challenges that many schools confront regarding environmental limitations to precollege engineering instruction. In this work, we present insights into the elicitation of non-functional requirements and the identification of constraints relevant for the design of learning technologies that can achieve greater ubiquity as well as improvement in the quality of user experience.

7.2   Problem Statement

In the realm of precollege engineering instruction, there is a need and opportunity
to expand the definition of learning technologies beyond Information and Communication Technologies (ICT). The customary definition of ICT in education refers
to products that allow the user to manage information electronically in a digital
form; for example, personal computers, Internet access, handset devices, microcontrollers and robot kits. However, learning technologies need not be limited to these
devices, rather many alternative platform possibilities are useful within engineering
education.

Engineering concepts and practices can also be cultivated with non-digital technology. This approach can be illustrated by looking at the engineering work created
during the Renaissance period (Ferguson et al. 1994) and more recently, through engineering projects presented by Sadler and collaborators (Sadler et al. 2000), which
demonstrate the feasibility of teaching engineering concepts with non-digital technology. To the best of our knowledge there is no research work which suggests that
ICT products are a necessity to ensure high quality precollege engineering education.

The integration of ICT into secondary schools is a complex task. At one time, it
was believed that it could be solved through policies which stipulate, for instance,
the number of computer labs and laptop carts with Internet access in schools. However, this strategy failed to reach many of the objectives that policy makers and
educators expected. Often, schools have computer labs but problems such as limited access, maintenance, and lack for training for teachers prevent full utilization
of these facilities (Cuban et al. 2001; Margolis et al. 2008).

Moreover, many countries are economically disadvantaged in terms of the acquisition of technologies. For example, Grace and Kenny (2003) state that if all of
Zimbabwes discretionary spending at the secondary level was used to provide students with computer access; this will buy 113 hours a year per student in front of the
computer. The remaining (approximately) 1240 hours per year would go unfunded.
In many of these areas, changes that would enable access to ICTs are considered
an unrealizable short-term goal. Rather, expanding the concept of learning technologies for precollege engineering education to include non-traditional learning
technologies would be particularly beneficial for such countries.

Many researchers have underscored the need to define the characteristics for successful learning technology planning and recognize requirements engineers as key
stakeholders in the integration of technologies into secondary schools (Fishman et
al. 2004). As advocates of precollege engineering we must remain cognizant of
challenges faced by teachers and students when utilizing learning technologies. In
general, we perceive a loose relationship between the designers of technologies and
the actual conditions under which teaching and learning take place in precollege settings. Consequently, the available technologies for engineering education are unable
to accommodate the needs of average teachers and students especially in resource-
constrained schools where access to ICTs is limited (Perraton and Creed 2002).
In order to reduce this disconnect, design guidelines are needed to enhance the

usability and ubiquity of learning technologies. Our work pursues the following objectives:

1. Illustrate quality requirements and constraints for the development and adaptation of learning technologies for precollege engineering instruction.
2. Establish a shared understanding regarding needs and effective practices for the design and development of learning technologies for precollege engineering instruction.
3. Develop a set of measures assessing engineering learning technologies respective to the schools where they are meant to employ.

To reach the aforementioned objectives, a goal-oriented engineering approach is proposed where the expectations for a product are defined as abstract goals and then refined into functional requirements, performance requirements, specific quality requirements and constraints (van Lamsweerde 2001).

Functional and performance requirements are fundamental to reach usability expectations and quality-in-use standards but they are specific to the system-to-be. Therefore, an attempt to render representations of these requirements to be used as general tenets by developers of learning technologies is impractical. However, we can suggest methods to identify quality requirements and constraints if we take into account the different types of schools that teaching and learning occur in, for example, schools with and without technology oriented classrooms i.e., classrooms with four or fewer computers (Margolis et al. 2008). Each category (or school cluster) is governed by similar physical and social constraints. Understanding these constraints is paramount for developers of new technologies and for individuals interested in adapting existing technologies to be employed in typical precollege classrooms.

The elicitation of requirements was performed through literature analysis and participant observation. Our literature analysis focused primarily on ICT. However, because technologies for engineering education are not restricted to ICT, participant observation was carried out to capture additional challenges that could not be inferred from the literature analysis. Our main purpose is to illustrate the usability issues related to technology-supported engineering learning at the precollege level.

7.3   Literature Analysis

To identify non-functional requirements we examined literature about challenges in using learning technologies in schools. The spate of published works on the topic is an indicator of its ongoing relevance. The majority of the papers did not consider engineering education specifically, but analyzed the integration of learning technologies in a broad manner from the perspective of teachers, students, and school systems. A majority of the publications focused on a narrow definition of ICT. Nevertheless, we consider relevant the findings on quality requirements and constraints presented in previous works because our experiences working with other types of

Table 7.1 Most frequently identified obstacles impeding the integration of technologies into schools

Identified Obstacles	Description	No. Refs	References
Technical Support	A reliable service which provides regular maintenance and repairs equipment on time	11	Beggs 2000; Blumenfeld et al. 2000; Cuban et al. 2001; Earle 2002; Fishman et al. 2004; Hew and Brush 2007; Groves and Zemel 2000; Kay et al. 2009; Keengwe and Onchwari 2008; Rogers 2000; Wilson et al. 2000
Availability and Accessibility	The presence of equipment in schools and the facility to use them with minimal restrictions	10	Beggs 2000; Blumenfeld et al. 2000; Cuban et al. 2001; Earle 2002; Fishman et al. 2004; Hew and Brush 2007; Groves and Zemel 2000; Hohlfeld et al. 2008; Kay et al. 2009; Rogers 2000
Teacher Training	Opportunities to receive instruction on how to use technology for pedagogical purposes	9	Blumenfeld et al. 2000; Cohen and Ball 2006; Earle 2002; Ertmer and Ottenbreit-Leftwich 2010; Fishman et al. 2004; Hew and Brush 2007; Kay et al. 2009; Keengwe and Onchwari 2008; Rogers 2000; Zhao and Frank 2003
Teacher Attitudes and Perceptions	How teachers feel about working with technology, Ranging from user experience to the motivation to consider and try new technologies	9	Beggs 2000; Blumenfeld et al. 2000; Earle 2002; Ertmer and Ottenbreit-Leftwich 2010; Hew and Brush 2007; Groves and Zemel 2000; Keengwe and Onchwari 2008; Rogers 2000; Zhao and Frank 2003
Learnability	How much time it will take to learn and use the technology effectively for pedagogical purposes	9	Beggs 2000; Cuban et al. 2001; Earle 2002; Fishman et al. 2004; Hew and Brush 2007; Groves and Zemel 2000; Kay et al. 2009; Keengwe and Onchwari 2008; Wilson et al. 2000
Integration of Curriculum and Technology	How much the technology aids the accomplishment of curriculum goals	8	Beggs 2000; Blumenfeld et al. 2000; Douglas et al. 2008; Earle 2002; Fishman et al. 2004; Groves and Zemel 2000; Keengwe and Onchwari 2008; Wilson et al. 2000
Administrative Support	Funding opportunities, reassurance and autonomy to use technology	8	Beggs 2000; Blumenfeld et al. 2000; Butler and Sellbom 2002; Earle 2002; Fishman et al. 2004; Hew and Brush 2007; Groves and Zemel 2000; Wilson et al. 2000
Ease of Use	Technology that is generally considered easy to learn and effective for pedagogical purposes	6	Beggs 2000; Butler and Sellbom 2002; Cohen and Ball 2006; Fishman et al. 2004; Groves and Zemel 2000; Wilson et al. 2000
Political Support	Endorsement and proactive Administrative policies that Support the use of technology in schools	5	Blumenfeld et al. 2000; Cohen and Ball 2006; Douglas et al. 2008; Fishman et al. 2004; Wilson et al. 2000
Quality of Available Technologies	Technologies that endure and satisfy the needs of school users	5	Blumenfeld et al. 2000; Cuban et al. 2001; Fishman et al. 2004; Hew and Brush 2007; Keengwe and Onchwari 2008

learning technologies such as robotics, microcontrollers and mechanical construction kits, suggest that these technologies struggle with barriers similar to those already identified in the current literature. Table 7.1 shows the top ten barriers founded in the analyzed publications from most frequently noted to least. Obviously, designers of learning technologies can have little to no influence on some of the identified barriers, such as political and administrative support. Yet the majority of the listed barriers should be considered at the products initial design phase.

7.4   Research Methodologies

We carried out an 18 hour intervention to expose teachers and students to a variety of learning technologies. Our aim was to observe and record participants perceptions about the use of technologies and the non-functional factors that might attract or detract participants from employing these technologies on a regular basis. Our work relies on three research methods: participant observation, student assessment and teacher interviews.

The study was performed with two groups of 7th grade students in an urban middle school in Tucson, AZ. We had no influence in dividing subjects into groups; the host school had already done this. Group A was composed of 22 students, while Group B contained 26 students. We devoted 9 one-hour sessions to work with each group. Students in both groups had average intellectual abilities and were diverse with respect to race, gender, socio-economic status, and achievement levels. The average age was 13 years old.

Both groups were taught engineering concepts and practices using different teaching methods commonly used in precollege education. Our motivation for using different teaching methods was not to render conclusions regarding the advantages of one method over the other but rather to reveal common issues linked to non-functional requirements regardless of the technology. Our intent was to employ learning technologies to teach concepts that span a spectrum of engineering tenets. The concepts of interest are (1) Systems, (2) Subsystems, (3) Process, (4) Control, (5) Feedback, (6) System inputs, (7) System Outputs, (8) Requirements Elicitation, (9) Optimization, and (10) Trade-offs. To foster a natural environment, each session was led by the same researcher and same participant teacher throughout the study. The host school provided a science classroom equipped with one computer for the teacher, three computers shared between all students, and an overhead projector.

7.4.1   Participant Observation

Participant observation is a contextual technique for the elicitation of system requirements. One of its advantages is that the researcher gains an insider view of the interactions in the location of interest (Jorgensen 1989). Additionally, participant

(a)   (b)   (c)

Fig. 7.2 Group A working with a variety of learning technologies, (a) eBlocks and integrated circuits on breadboards, (b) LEGO Mindstorms robot, (c) Ferris wheel built with KNEX

observation exposes aspects of ordinary use that are difficult to capture using traditional requirement elicitation techniques and can reveal a more empathetic view of the users experience (Wright and McCarthy 2008). Our observation was overt (i.e., the participants were aware of the researcher role within the natural setting) and detailed field notes were recorded at the end of each one-hour session.

The research team collected data using quantitative and qualitative methods. The main advantage of qualitative methods is that they force the researcher to delve into the complexity of the problem rather than abstract it away (Seaman 1999). While the benefits of qualitative methods to elicit quality requirements and constraints are well defined, results can be difficult to summarize and are frequently considered softer or fuzzier (Seaman 1999). In contrast, quantitative techniques provide researchers with hard or precise results but can fail to capture the details of a phenomenon. While our study relies heavily on qualitative research, we employed quantitative methods to explain some aspects of our findings. Objective and subjective measures of usability are valuable for researchers particularly when researchers have to interpret, compare and relate data, as these different evaluation methods could suggest inconsistent results (Creswell and Plano Clark 2011; Hornbaek 2006).

Group A was presented with a conventional teaching method, in which each session focused on the presentation of an engineering concept. Then the concept was reinforced either through a learning activity or through a demonstration provided by the participant teacher or researcher . The learning activities required students to interact with one another through a hands-on task while following cookbook style instructions (Fig.7.2). The technologies used with Group A were: 1) integrated circuits and electric components on breadboards* (Logic gates, LEDs, resistors, and push buttons), 2) eBlocks* (Phalke and Lysecky 2010), 3) the MicroBug BEAM bot (Velleman 2011), 4) LEGO Mindstorms (Lego 2011), 5) KNEX* construction kits (KNEX 2012). Additionally, each student in Group A participated in a writing competition in which they had to provide a solution to an engineering problem that required formal reasoning about the technical concepts and practices introduced during the intervention period.

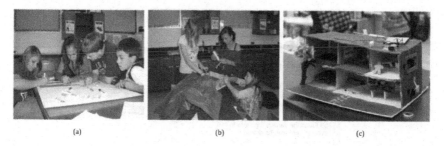

Fig. 7.3 Group B working in a long-term competition project, (a) from drafting a plan to construct a model to (b) construction of a final smart home enhanced with sensors and actuators

A long-term competition project was used with students in Group B, where students, in groups of four to five, built a scale model of a smart-home for a hypothetical family (Fig.7.3). Each team was provided with a story line that described the family members and their respective life styles. Teams were also provided with an initial amount of tokens, which could be exchanged for construction material or electrical components for use in their models. Teams could win extra tokens during the intervention by completing mini-challenges such as correctly answering questions in a quiz covering the engineering topics taught in previous sessions. Using standard sensors and actuators, battery- and solar-powered lights, and stepper motors, students had the opportunity to build small electronic systems within their smart home models. At the end of the intervention students participated in a science fair style presentation and were evaluated on the basis of creativity, aesthetics, cost (number of tokens spent in each project), and functionality.

7.4.2 Assessment of Engineering Concepts

An assessment questionnaire composed of ten open-ended questions was developed to evaluate participants engineering knowledge. We were of course interested in whether these learning technologies helped student learning, as well as usability and accessibility issues. However, due to absences, not all the participants were able to accomplish both pre- and post-assessments. Our findings only include data obtained from subjects that completed at least 80% of the sessions.

7.4.3 Teacher Interviews

To understand critical quality requirements such as usability and user experience issues (UX) a series of interviews with the participant teacher were conducted

Rating scale: 1 – Strongly disagree to 5 – Strongly agree

System Usability Scale (SUS)

1.I think that I would like to use this system   12345
frequently

2.I found the system unnecessarily   12345
complex

3.I thought the system was easy to use   12345

4.I think that I would need the support of a   12345
technical person to be able to use this
system

5.I found the various functions in this   12345
system were well integrated

Fig. 7.4 Sample of items included in the SUS scale

throughout the intervention period. Usability encompasses how easily a technology can be employed and how appropriate it is for a particular purpose (Brooke 1996). User experience is an aspect of usability and includes factors such as affect, emotion, fun, aesthetics, and flow (Law 2011). Traditionally usability is measured quantitatively through scaled instruments while UX requires qualitative research. Therefore, we employed two research methods to collect data for both constructs: usability and UX. For the former we conducted a structured interview based on the items of a widely used and validated scale. For the latter, we used an unstructured interview.

The participant teacher has worked at the middle school level for 20 years, 18 as a science teacher. Moreover, she serves as the schools Science Facilitator and is responsible for training incoming science teachers and disseminating the state-of-the-art in science education within her school. The teacher identified herself as a digital immigrant, a term applied to describe subjects who did not grow up in a ubiquitous digital environment and who have learned skills to adapt to technologies broadly used and often considered essential by digital natives. Immigrants of different cultures adapt to new cultures at different paces and degrees, but usually retain traits from their old cultures, for example, an accent. Likewise, digital immigrants show traits that distinguish them from natives including how much they take advantage of technologies in their daily lives and their job duties (Prensky 2001). However, we urge caution in indiscriminately using the terms digital immigrants to describe teachers and digital natives to describe students, as recent studies suggest these assumptions could lead to stereotypes with negative implications for education (Bennet et al. 2009).

The participant teacher responded to a brief set of questions at the end of each session regarding the usability of the technologies employed during the session. This set of interviews was based on the items of the System Usability Scale (SUS) (Brooke 1996). SUS is a technology independent usability survey scale for evaluating a wide variety of software or hardware systems. It consists of ten items and provides a reference score that can be mapped to adjective ratings and acceptability categories (Bangor et al. 2008). The SUS scale was chosen because its items have been assessed for reliability and validity (Bangor et al. 2008; Brooke 1996). Fig.7.4 provides a sample of items included in the SUS scale.

Research instruments consisting of close-ended items such as the SUS scale can be very effective when comparing two prototypes of the same invention. However when close-ended items are used to compare satisfaction with different inventions, we recommend following every item in the instrument with an open question that prompts the respondent to elaborate on their response. This practice lets the requirements elicitor learn about factors that influence the context of use. For example, item 3 in the SUS scale (Fig. 4) seeks to measure Ease of us. However, simply providing a number is not enough to understand which factors influenced why a particular score was given for the evaluated technology.

A final unstructured interview with the participant teacher was carried out to better understand UX aspects not covered in the SUS usability scale. These UX aspects will help to define parameters for a general model to develop usable learning technologies for a wide range of teachers and students. While our example comes from a middle school in Tucson, the outcomes are not intended to be directly applicable for all public middle schools in the United States. Rather, they suggest how the quality model provided in Section 6 can serve as a general guiding tool for the effective development of high quality, accessible learning technologies.

7.5   Findings

7.5.1   Outcomes of Assessment of Engineering Concepts

To draw any conclusions regarding the requirements of learning technologies, it is critical to establish that students actually learn during the intervention. To that end, dependent t-tests of pre- and post-assessments were performed to determine the significance of learning in each group. As shown in Table 7.2, both groups showed a significant difference between pre- and post- assessments.

Table 7.2   Significance of learning outcomes in Group A and Group B

Group A		
Pre-assessments	M = 2.16	SD = 1.39
Post-assessments	M = 3.75	SD = 2.04
t(13) = -4.37		
p 0.001		
Group B		
Pre-assessments	M = 1.46	SD = 1.13
Post-assessments	M = 3.42	SD = 1.18
t(18) = - 4.76		
p = 0.005		

The statistical analysis suggests that participants were able to learn basic engineering concepts both through using a traditional methodology (Group A) and through participating in a long-term project (Group B). In general, higher quality responses were obtained from Group B in that a greater percentage of members of Group B provided more sophisticated conceptual answers than Group A. However, an independent t-study of the quality of answers provided by both groups showed no significant differences.

7.5.2 Outcomes of Structured Interviews

Table7.3 shows the adjective ratings provided by the participating teacher to each of the learning technologies utilized during the intervention. The adjective-based ratings are calculated from the teachers answers to the SUS scale (Bangor et al. 2008). Since the power of the SUS scale relies on the number of subjects answering the questions, a limitation of our study is that only one subject, the participant teacher, answered the SUS questionnaire. Therefore, the presented scores should not be used to infer advantages of one technology over another. Nonetheless, the results served as a measure of the technologies usability within the scope of our intervention. After mapping the SUS scale scores to their corresponding rating adjectives, the results show that KNEX (2012), a construction set consisting of bricks, rods, wheels, gears and connectors, was perceived as an excellent learning technology (Fig. 2c). We prompted the teacher to elaborate on her decision to rate this particular technology higher than the rest. She was very clear in pointing out that even though the mapping of scores to rating adjectives show an Excellent result, she would not consider the technology as Excellent. In retrospect she felt her higher ratings for KNEX could have been influenced by the simplicity of the product and the fact that she saw several ways to use it for teaching purposes. She also articulated her concern that although KNEX was easy to use, it might be limited to teaching simple concepts in short-term projects. She expressed doubt that it would be an effective tool for teaching complex engineering concepts.

Table 7.3 Usability SUS scores and adjective ratings for engineering learning technologies

Technology	SUS Score	Rating Adjective
Integrated Circuits and Breadboards	13.5	Worst Imaginable
eBlocks	51.7	Acceptable
Microbug	31.5	Poor/ Worst Imaginable
Lego Mindstorms	54	Good
K'NEX	76.5	Excellent
Standard Sensors, Actuators and Power Sources	54	Good

In contrast, the participant teacher was very interested in the Lego Mindstorms (2011), one of the most popular robotics kits for young users (Fig. 2b). However, the teacher felt she would need the support of a technical person to learn how to use the technology as well as needing more access to the schools computer lab. She also expressed doubt that the monetary resources to acquire enough Mindstorms kits for her class were available in her school.

The eBlocks educational kits (Phalke and Lysecky 2010), shown in Fig. 2a, are composed of a set of fixed function blocks that enable non-expert construction of embedded system applications. In this case the teacher was satisfied with how easy the eBlocks were to learn for herself and the students. Her main concern was with the robustness of the current prototype, as some of the eBlocks broke while students where using them and there was no easy way to repair them on-site.

Standard sensors, actuators and power sources were rated higher than we expected. The teacher perceived it advantageous to expose students to standard electrical components such as push buttons, switches, potentiometers, servomotors, LEDs, solar cells and batteries. Additionally, these components could be adapted so the students could connect them using only crimp pliers and butt connectors as opposed to a soldering gun.

The MicroBug (Vellman Inc 2011) is an example of BEAM bots used by hobbyists and sometimes educators. BEAM bots are composed of low-cost electronic components, so their acquisition cost is lower than most robotic alternatives. However, two disadvantages of using BEAM bots are that the components usually need to be soldered together, and once the final product is built, it is difficult to take the robot apart to reuse its components in later projects. The teacher commented that the high supervision required when students are using soldering guns or similar tools could not be provided in a regular day. She also felt that since she could only supervise a small number of students at one time while working with the MicroBug, employing the technology could easily lead to classroom management problems.

The use of integrated circuits and breadboards was the least popular for both the teacher and students. The teachers frustration while working with integrated circuits and breadboards was clear during the intervention. For example, when students were unable to follow basic instructions to use AND and OR gates to turn LEDs on and off, they lost their engagement with the activity, and were unable to keep themselves on-task. In this case, the teachers negative score can be attributed to the students response to the activity.

7.5.3   Outcomes of Unstructured Interview

The main objective of the unstructured interview was to understand the compromises a teacher faces when deciding to acquire a dedicated learning technology or develop their own lesson plans with other technologies not specifically designed for

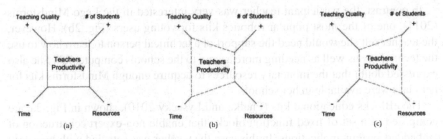

Fig. 7.5 Explaining teachers productivity in precollege engineering education through the Devils square. (a) A compromise scenario between competitive factors (b) Teachers perception of productivity using learning technologies (b) Teachers perception of productivity using non-dedicated technologies

education purposes (non-dedicated technologies). As expected, several constraints elicited from the teacher have already been identified in previous works (Table7.1), specifically, (1) limited access to the computer lab, (2) funding, (3) quality of technical support, (4) ongoing education, and (5) basic engineering education.

Limited access to the computer lab. Like other teachers in the school (35 teachers in all) the participant teacher has to reserve the computer lab every time she wants to use it for her class. The result is that she can only use the computer lab 2 days per school quarter (a 45 day period), not enough time for students to work on long-term engineering projects.

Funding. Funding to purchase any type of technology is scarce and competitive. Though several mini-grant programs are available, writing these grants requires considerable time outside of school hours. Not surprisingly, teachers are easily discouraged from submission when one or a few of their applications are denied.

Quality of technical support. Teachers at the host school generally feel deserted with regard to technical support. The participant teacher shared with us that in her school it sometimes takes over a month to get technical help. Therefore when technical issues arise, teachers have little recourse but to try to figure out and fix the problems themselves.

Accessible Training. Teachers also experience as a challenge access to training in using new learning technologies. Though technology vendors usually offer workshops and courses at special rates to practitioner teachers, these rates are typically be covered by the teacher instead of the school they work for.

Basic Engineering Education. Curriculum is certainly available for many commercial products; however there is a need for open-source engineering curriculum that describes engineering concepts in age-appropriate ways apart from a particular learning technology. Such materials would enable teachers (and students) to understand engineering principles at a higher level, and to relate engineering to other disciplines such as science and math. In addition, such materials would enable teachers to feel more comfortable at designing their own lessons using available resources.

Regardless of the challenges identified, teachers in the host school are largely interested in using learning technologies in their classrooms. The salient point of the interview was the teachers reasoning for making decisions about acquiring a learning technology. She was highly influenced by what she perceives as teacher productivity, defined as the ability to teach quality concepts in a shorter time; given the resources the teacher has access to in her school. Success in using learning technologies in precollege engineering education can be seen as a trade-off between four key factors: 1) the percentage of students within the classroom that receive hands-on-experience with the technology. For example, when a limited number of computers are provided, some students tend to own the technology while others passively observe them or engage in other non-productive activities; 2) the resources required to use the learning technology, including acquisition cost, physical space, technical support, need of computers or Internet access, etc.; 3) the quality of teaching enabled by the technology, evaluated by the breadth and depth of concepts that can be taught, and 4) the time beyond paid work hours teachers need to invest to prepare a teaching session.

The teachers reasoning is explained by using a model known as the Devils Square, which was originally developed to understand the productivity of software engineers (Sneed 1989). We have adapted this model to explain the teachers perceived productivity during the intervention period (Fig.7.5a). The four corners of the Devils square represent desired but competitive factors that had to be balanced by the instructor while using learning technologies. In principle, it is desirable to maximize the factors positioned at the squares upper corners, while minimizing the factors positioned at the lower corners. In a balanced scenario, the teachers productivity is represented as an inner square; this means that the four competitive factors have been evenly balanced. The teachers productivity is considered a constant, meaning that it will always cover the same area inside the outer square regardless of whether its shape is a square, a parallelogram, or a triangle. The shape of the inner figure depends on how the teacher finds a balance between the identified competitor factors.

During the intervention, we worked with dedicated learning technologies (Group A) as well as built a lesson plan which relies on technologies not specifically designed for educational purposes (Group B). Dedicated learning technologies (e.g. Mindstorms) minimize the preparation time teachers need to invest in planning a lesson (these products are regularly accompanied by rich curriculum) and maximize the quality of learning but require significant resources. Therefore, with the exception of technology-rich schools, only a small percentage of students will have the opportunity to work directly with the technology (Fig.7.5b).

In comparison, developing engineering lessons with technology not specifically designed for educational purposes (e.g. standard sensors, actuators and power sources) can be time consuming, as many of these technologies have to be adapted by the teachers before these technologies can be used in the classroom. Additionally teachers have to invest time crafting effective lessons plans (Fig.7.5c). On the other hand, non-dedicated learning technologies can be acquired in high volume at a modest price, allowing more students to have hands-on experiences. However, the

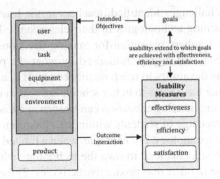

Fig. 7.6 ISO 9241-11 (1998) Usability Framework

quality of teaching and learning can be limited by the affordances of non-dedicated technology i.e., the strengths and weaknesses of technologies with respect to the possibilities they offer the people that might use them (Gaver 1991). Moreover non-dedicated technologies require a significant cognitive load from teachers because the effectiveness of these technologies are highly dependent on the teachers ability to use them in a constructive way.

Clearly the optimum scenario entails teachers using technologies that maximize their teaching quality and allow significant numbers of students to have hands-on experience, while minimizing the time teachers must invest after hours in preparing curriculum and remaining economically feasible for a given school.

7.6   A Basic Model for the Development and Adaptation of Learning Technologies

Considering the precollege community as a uniform market for learning technologies is an unsustainable notion. There are inequalities in the level of constraints, ICT, and resources available. While reports from institutions such as the National Center of Education Statistics (NCES) (Gary et al. 2008) are useful to keep track of progress relating to the distribution of technology at a macro-level, they can be misleading if interpreted as an indicator of ICT integration at the meso- and micro-levels of educational settings. In our experience, the use of ICT resources at our host school differed considerable from what the NCES data may have led one to expect.

7.6.1   Quality Requirements

General guiding tools are needed to capture quality requirements in the domain of learning technologies. An adapted version of the usability model proposed by the

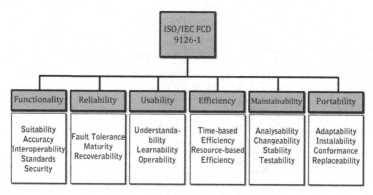

Fig. 7.7 The ISO/IEC FCD 9126-1 (1998) quality model to evaluate quality-in-use

International Organization for Standardization, ISO 9241-11 (1998) (Fig.7.6), and the quality model included in ISO/IE 9126-1 (1998) (Fig.7.7) can serve to enumerate the requirements sought in learning technologies by stakeholders.

The ISO 9241-11 usability framework was developed to define usability for hardware-software systems. It strives to facilitate the design not only of technical functional products, but also products that are usable by consumers. The framework comprises three main elements: a description of intended goals, usability measures such as effectiveness, efficiency and satisfaction, and a description of the context of use, which encompasses user expertise, tasks necessary to achieve a goal, available equipment, and the social and physical environment where the system will be used. Effectiveness is a measure of how well users achieve specified goals and reflects the resources expended in meeting the user specified goals, while satisfaction addresses user attitudes towards the product.

The literature analysis and participant observation suggest that there exist clusters of schools, which share goals and operate in contexts that lead to similar usability challenges. For example, most of the research literature cited in Section 3 focuses on contexts within the US school system and that of other high-income countries. Schools that share similar characteristics may be able to directly apply the findings of these works. However, the findings may not be applicable for resource-constrained schools. Likewise, there could be different school clusters coexisting in a country, for example clusters can be defined by community type such as urban versus rural. Educational policies also play a role in dividing schools clusters as national and local standards may differ from country to country and state to state. Other factors to take into account are social, political and cultural norms. In sum, schools within a given cluster or category share fundamental structural factors and have common issues regarding learning environment; examples of these factors are provided in Fig.7.8.

Rather than designing products for pre-college engineering instruction with a single model in mind, we believe that educational products will be more successful if the wide variety of school characteristics is taken into account in the early planning

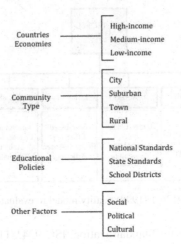

Fig. 7.8 Factors that can determine school clusters

stages. The result will be customized versions of general quality models, based on the target school cluster, that ensure the inheritance of crucial attributes from the parent model.

Many of the quality requirements elicited during our research align with the requirements included in the ISO 9126-1 quality model (Fig. 7.7). While the ISO 9126-1 is a general quality framework intended for evaluating software systems, it does not prescribe specific requirements, which makes it malleable enough to be used in other domains through applying a customization process (Behkamal et al. 2009; Chua and Dyson 2004). The customization process is key to successfully using general quality models in a specific domain. Our model provides the means to facilitate customization by analyzing the factors or parameters that affect the usability and UX of students and teachers using learning technologies for engineering education.

7.6.2 Determining Quality-in-Use of Engineering Learning Technologies

We have chosen the ISO 9126-1 as an example of a general model because of its high recognition, adaptability and validity. The ISO 9126-1 (1998) quality model comprises by six product characteristics that are further divided into subcategories. Functionality describes needs of the platform. Reliability relates to the capability of the product to maintain its performance over time. Usability evaluates the effort required by the end user or set of end users. Efficiency relates to the relationship between the performance of the product and the amount of the resources used to obtain the desired performance. Maintainability describes the effort needed to modify

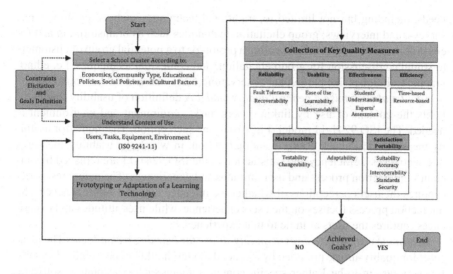

Fig. 7.9 Basic model for the development and adaptation of learning technologies

the platform. Portability indicates the ability of the platform to move across environments.

Learning technology products that meet these quality requirements should facilitate quality-in-use for stakeholders. Both teachers and students benefit from the functionality, reliability, usability and efficiency of the product. Technical support benefits from the maintainability requirements. The school benefits from the product portability requirement since, as the shared resource can move across a variety of classrooms without dedicated ICT requirements. Using a customized model based on the ISO 9241-11 and ISO/IE 9126-1 as a guiding tool to achieve usability expectations and quality-in-use can facilitate adoption of engineering instruction technologies (Fig.7.9).

Identify representative school cluster

The avenue to assessing engineering learning technologies starts by tackling the barriers of usability, adoption and sustainability at the classroom and school level. Schools can be categorized in terms of their goals, needs, and constraints. Any one size fits all product will inevitably preclude some school clusters from acquiring the technology and in other cases schools might be prevented from taking full advantage of their ICT resources. Instead, we believe that our goal oriented requirements engineering approach, which acknowledges the differences in constraints and quality requirements among school types, is a more promising approach.

Apply goal oriented requirements engineering elicitation techniques

Once a cohesive school cluster has been recognized, it is imperative for the requirements engineer to identify the stakeholders and their needs in relation to the system-to-be. Common stakeholders are teachers, students, administrators, technical personnel and political and social organizations (e.g. school board). Numerous methodologies are available for the elicitations of requirements and stakeholders needs, including but not limited to, traditional techniques such as questionnaires, surveys, and interviews; group elicitation techniques such as brainstorming and focus groups; prototyping, or: presenting a prototype to a potential group of customers to stimulate discussion for group elicitation; and contextual techniques, i.e. ethnographic techniques like participant observation (Nuseibeh and Easterbrook 2000).

Adopt and customize a general quality model A definition of usability that highlights the context of use by linking users, tasks, equipment and environment is needed. The ISO 9241-11 model (Fig.7.6) is one such option. The model for usability measures (Hornbaek 2006) is another option, in which the usability measures of effectiveness, efficiency and satisfaction in the ISO 9241-11 are replaced by outcomes, interaction process and user attitudes and experiences. The outcomes aspect captures the users perceptions of whether the intended outcomes were reached.The interaction process focuses on the users experience, while user attitudes and experiences captures the users attitude to that experience.

Determine critical measures of quality-in-use for learning technologies The measures for quality-in-use provided by a general model should not be weighted equally for every system-to-be. Rather, specific content domains such as accounting software, programming environments, domestic appliances, and learning technologies place emphasis on different aspects of quality-in-use. Methods are available to rate the importance of the individual quality metrics when general models are applied to specific domains (Behkamal et al. 2009). A common issue associated with the use of general quality models across various domains lies in the interpretation of the measures. For example, in the education domain, high task-completion times can be indicators of students engagement and motivation while in domains where worker productivity is crucial, high task-completion times might be unacceptable (Behkamal et al. 2009). It is critical for the requirements engineers to adequately define how each measure will be construed to avoid interpretation errors during the design and evaluation process.

This paper provides a fundamental but not exhaustive list of measures to determine quality-in-use of learning technologies for engineering education. These measure are categorized as measures of reliability, usability, effectiveness, efficiency, maintainability, portability and satisfaction.

Measures of Reliability address the ability of the learning technology to keep functioning consistently as expected over time. Fault Tolerance reflects the robustness of a learning technology across different levels of usage, indicated by the average time that takes a technology to fail. Recoverability captures a technologys ability to be restored to a functional state preferably the same state as before the failure occurred and evaluates both the recovery time needed to restore the system as well as the recovered state.

Measures of Usability addresses how easily a technology can be employed and how appropriate it is for a given purpose. Specifically, Ease of use considers cognitive challenges independent of the subject matter, emphasizing how important it is for learning technologies to eliminate details and steps that prevent users from learning the core subject contents. Papert (1980) and Resnick et al (2009) explained ease of use in the domain of programming languages for children as having low floors the trait of a programming language that allows users to get started programming algorithms easily. Equally important are the traits of high ceilings and wide walls, referring to the technologys affordance to allow its users to increase the complexity of end products, provide opportunities for exploring a diversity of topics, and accommodate different learning styles. Developers can gauge the ease of use of a particular learning technology through direct observation, thinking-out-loud techniques, interviews and retrospective questionnaires. Learnability refers to the ability of the system-to-be to provide resources for teachers and students that are perceived as a learning on your own experience. For example, in our intervention, the teachers interest in the Lego Mindstorms was tempered by her perception that she would need to attend a specialized course to use the technology. A variety of collection methods such as interviews, direct observation and timing techniques can help to gauge the learnability of the platform. Lastly,

Understandability refers to the necessity of a learning technology being clear to the users and designed to prevent mistakes. Usability can be described quantitatively as the number of affordances used by students and teachers divided by the total number of affordances offered by a given learning technology.

Measures of Effectiveness determine if the employment of a technology facilitates the learning of engineering concepts and practices. First, Measures of student understanding indicate if the system facilitates the students conceptual and procedural understanding of the topic domain. For the assessment of procedural knowledge, retrospective interviews and thinking-out-loud approaches are recommended. The assessment of conceptual knowledge can be achieved by analyzing the quality of students answers to open-ended questions. Next, an Expert assessment helps to evaluate student artifacts produced as a result of this learning technology (Hornbaek 2006). In the intervention presented in previous sections, engineering faculty and graduate students were recruited to evaluate the end products developed by students essays and smart home models. The engineering experts who assessed the artifacts also served as a knowledge source to improve the technology in accordance with its educational goals.

Measures of Efficiency evaluate how human intellectual and physical resources are spent to achieve a meaningful understanding of the content domain. Time-based Efficiency can be evaluated simply by recording the set-up time to use the learning technology in the classroom and the time needed to clear the working space. In contrast, Resource-based efficiency assesses the ability of the targeted school cluster to take advantage of their ICT resources. The number of students concurrently interacting with a given technology without a teachers direct supervision is also a good measure of this metric.

Measures of Maintainability capture the effort required to diagnose and troubleshoot a learning technology on-site. Testability enables users to verify if the system is working appropriately and provides meaningful status cues when the technology is failing. Interacting with users and recording the interpretability of the cues provided by the system when in failure or non-optimal status is a good measure of this metric. Changeability denotes the ability of a user to modify the technology to better support their learning goals in an effortless and cost-effective way.

Observing a user modify or replace one aspect of a learning technology is one method to evaluate the changeability of a given platform.

Measurements of Portability are paramount in educational settings, especially for schools constrained by inadequate ICT resources or classroom space. The ability to use the learning technology in a diversity of learning environments facilitates the scalability of the product to wider audiences. Adaptability refers to how the learning technology can be changed to satisfy constraints imposed by different learning environments. It can be assessed by dividing the number of educational experiences possible in a particular learning environment by the number of experiences possible in an ideal environment.

Measures of Satisfaction reflect the perceptions of teachers and students towards using a technology. Satisfaction relates to the users experiences and how the outcomes of such experiences are recalled. Satisfaction is usually assessed through questionnaires, attitude rating scales, ranking tables and open interviews. Preference measures should be used with both teachers and students, since they might have different needs and priorities. Measuring the Perception of Outcomes is also crucial to demonstrating the learning advantages provided by a given technology. Improvements in knowledge and understanding need to be obvious to the teachers and students, and based on quantitative evidence so that the investment of resources can be justified.

7.7 Discussion

The experience of working alongside a teacher in a typical school environment has enabled a better understanding of the challenges faced by teachers using a variety of learning technologies. The path to assessing the usability of engineering learning technologies is long and beset with challenges. Achieving technical functionality is not enough; we must seek out solutions with broad applicability across a variety of learning environments. While designing high-end products has benefits, there remain many obstacles in obtaining these technologies and effectively utilizing them. We propose tackling the problem by first developing notions of school clusters, characterized by similarities in available resources and constraints. Second, we propose adopting a requirements engineering approach that focuses on quality-in-use, employing general usability and quality models with key usability and quality measures specific to the domain of precollege engineering education. The proposed model is by no means complete and mature, but is a step towards requirements specifications that support the development of more inclusive learning technologies for precollege engineering education. To further corroborate the suggested strategies, future research might focus on different types of school clusters as well as further evaluation of needs identified within the same cluster. We look forward to further research that strengthens, refines, and expands this work.

References

1. Bangor, A., Kortum, P., Miller, J.: An Empirical Evaluation of the System Usability Scale. Journal of Human-Computer Interaction 24(6), 574–594 (2008)
2. Beggs, T.A.: Influences and Barriers to the Adoption of Instructional Technology. In: Proceedings of the Mid-South Instructional Technology Conference, Murfreesboro, TN (2000)
3. Behkamal, B., Kahani, M., Kazem Akbari, M.: Customizing ISO 9126 quality model for evaluation of B2B applications. Information and Software Technology (51), 599-609 (2009)
4. Bennet, S., Maton, K., Kermit, L.: The 'digital natives' debate: a critical review of the evidence. British Journal of Educational Technology (39), 775–786 (2008)
5. Bevan, N.: Measuring usability as quality of use. Software Quality Journal (4), 115–150 (1995)
6. Bevan, N.: Quality in Use: Meeting User Needs for Quality. Journal of Systems and Software (49), 89–96 (1999)
7. Blumenfeld, P., Fishman, B., Krajcik, J., Marx, R.W., Solloway, E.: Creating Usable Innovations in Systemic Reform: Scaling Up Technology-Embedded Project-Based Science in Urban Schools. Educational Psychology 35(3), 149–164 (2000)
8. Brooke, J.: SUS: A 'quick and dirty' usability scale, Usability Evaluation in Industry. Taylor and Francis, London (1996)
9. Butler, D., Sellbom, M.: Barriers to Adoption Technology for Teaching and Learning. Education Quarterly 25(2), 22–28 (2002)
10. Robotics Academy, Carnegie Mellon Robotics Academy (2010), http://www.education.rec.ri.cmu.edu
11. Chua, B., Dyson, L.: Applying the ISO9126 model to the evaluation of an e-learning system. In: Proceedings of the Australian Society for Computers in Learning Tertiary Education (ASCILITE) Conference, Perth, Australia, pp. 184–190 (2004)
12. Cohen, D., Ball, D.: Educational innovation and the problem of scale. In: Scale-Up In Education: Ideas In Principle. Rowman and Littlefield Publishers, Lanham (2006)
13. National Research Council, A Framework for K-12 Science Education: Practices, Cross-cutting Concepts, and Core Ideas. The National Academies Press, Washington, DC (2011)
14. Creswell, J.W., Plano Clark, V.L.: Designing and Conducting Mixed Methods Research, 2nd edn. SAGE Publications Inc., Thousand Oaks (2011)
15. Cuban, L., Kirkpatrick, H., Peck, C.: High Access and Low Use of Technologies in High School Classrooms: Explaining an Apparent Paradox. American Educational Research Journal 38(4), 813–834 (2001)
16. Douglas, S., Christensen, M., Orsak, G.: Designing Pre-College Engineering Curricula and Technology: Lessons Learned from the Infinity Project. Proceedings of the IEEE 96(6), 1035–1048 (2008)
17. Earle, R.: The Integration of Instructional Technology into Public Education: Promises and Challenges. ET Magazine 42(1), 5–13 (2002), http://bookstoread.com/etp/earle.pdf (retrieved June 24, 2012)
18. Ertmer, P., Ottenbreit-Leftwich, A.: Teacher Technology Change: How Knowledge, Confidence, Beliefs and Culture Intersect. Journal of Research on Technology in Education 42(3), 255–284 (2010)
19. Ferguson, E.: Engineering and the Mind's Eye. MIT Press (1994)
20. Fishman, B., Marx, R., Blumenfeld, P., Krajcik, J.: Creating a Framework for Research on Systematic Technology Innovations. The Journal of the Learning Sciences 13(1), 43–76 (2004)

21. Hew, K.F., Brush, T.: Integrating Technology into K-12 Teaching and Learning: Current Knowledge Gaps and Recommendations for Future Research. Educational Technology Research Development (55), 223–252 (2007)
22. Foster, P.N.: The Relationship Among Science, Technology and Engineering in K-12 Education. Connecticut Journal of Science Education (42), 48–53 (2005)
23. Gary, L., Thomas, N., Lewis, L.: Educational Technologies in USA public schools: Fall 2008. U.S. Dept. of Education, National Center for Education Statistics, NCES 2010-034 (2008)
24. Gaver, W.: Technology Affordances. In: Proceedings of the Conference on Human Factors in Computing Systems, New Orleans, LA, pp. 79–84 (1991)
25. Glinz, M.: On Non-Functional Requirements. In: Proceedings of the IEEE International Requirements Engineering Conference, New Delhi, India, pp. 21–26 (2007)
26. Grace, J., Kenny, C.: A short review of information and communication technologies and basic education in LDCs- what is useful, what is sustainable? International Journal of Educational Development (23), 627–636 (2003)
27. Groves, M., Zemel, P.C.: Instructional Technology Adoption in Higher Education: an Action Research Case Study. International Journal of Instructional Media 27(1), 57–65 (2000)
28. Hohlfeld, T., Ritzhaupt, A., Barron, A.E., Kemker, K.: Examining the Digital Divide in K-12 Public Schools: Four-year Trends for Supporting ITC Literacy in Florida. Computers and Education (51), 1648–1663 (2008)
29. Hornbaek, K.: Current practice in measuring usability: Challenges to usability studies and research. International Journal of Human Computer Studies (64 2, 79–102 (2006)
30. ISO/IEC, F.C.D.: 9126-1, Software product quality - Part 1: Quality model (1998)
31. ISO 9241-11, Ergonomic requirements for office work with visual display terminals (VDTs) - Part 11 Guidance on usability (1998)
32. Jorgensen, D.L.: Participant Observation: A Methodology for Human Studies. Sage Publications, Thousand Oaks (1989)
33. Katehi, L., Pearson, G., Feder, M.: Engineering in K-12 Education, Understanding the Status and Improving the Prospects. The National Academies Press, Washington D.C. (2009)
34. Kay, R., Knaack, L., Petrarca, D.: Exploring Teachers Perceptions of Web-based Learning Tools. Interdisciplinary Journal of E-Learning and Learning Objects (5), 27–50 (2009)
35. Keengwe, J., Onchwari, G.: Computer Technology Integration and Student Learning: Barriers and Promise. Journal of Science Education and Technology 17(6), 560–565 (2008)
36. K'NEX, The World's Most Creative Construction and Building Toys (2012), http://www.knex.com
37. Law, E.L.C.: The Measurability and Predictability of User Experience. In: Proceedings of the Symposium on Engineering Interactive Computing Systems, Pisa, Italy, pp. 1–10 (2011)
38. van Lamsweerde, A.: Goal Oriented Requirements Engineering: A Guide Tour. In: Proceedings of the IEEE International Symposium on Requirements Engineering, Toronto, Canada, pp. 249–261 (2001)
39. Lawson, A.: Teaching Inquiry Science in Middle and Secondary Schools. Sage Publications, Thousand Oaks (2010)
40. LEGO, LEGO Mindstorms (2011), http://mindstorms.lego.com
41. Lu, S.C.Y.: Collective rationality of group decisions in collaborative engineering. International Journal of Collaborative Engineering 1(1-2), 38–74 (2009)

42. Margolis, J., Estrella, R., Goode, J., Jellison Holme, J., Nao, K.: Stuck in the Shallow End: Education, Race, and Computing. The MIT Press, Cambridge (2008)
43. Nuseibeh, B., Easterbrook, S.: Requirements engineering: a roadmap. In: Proceedings of the International Conference on Software Engineering, Limerick, Ireland, pp. 35–46 (2000)
44. Papert, S.: Mindstorms: Children, Computers, and Powerful Ideas. Basic Books, New York (1980)
45. Perraton, H., Creed, C.: Applying new technologies and cost-effective delivery systems in basic education. Education for All Secretariat, UNESCO, Paris (2002)
46. Peske, H.G., Haycock, K.: Teaching inequality: How poor and minority students are shortchanged on teacher quality. The Education Trust, Washington, D.C. (2006)
47. Phalke, A., Lysecky, S.: Adapting the eBlock Platform for Middle School STEM Projects: Initial Platform Usability Testing. IEEE Transaction on Learning Technologies 3(2), 152–164 (2010)
48. Prensky, M.: Digital natives, digital immigrants. On the Horizon 9(5), 1–6 (2001)
49. Resnick, M., Maloney, J., Monroy-Hernández, A., Rusk, N., Eastmond, E., Brennan, K., Millner, A., Rosenbaum, E., Silver, J., Silverman, B., Kafai, Y.: Scratch: Programming for all. Communications of the ACM 52(11), 60–67 (2009)
50. Riojas, M., Lysecky, S., Rozenblit, J.: Educational Technologies for Precollege Engineering Education. IEEE Transactions on Learning Technologies 5(1), 20–37 (2012)
51. Rogers, P.L.: Barriers to Adopting Emerging Technologies in Education. Journal of Educational Computing Research 22(4), 455–472 (2000)
52. Sadler, P., Coyle, H., Schwartz, M.: Engineering Competitions in Middle School Classroom. Journal of the Learning Sciences 9(3), 299–327 (2000)
53. Seaman, C.B.: Qualitative Methods in Empirical Studies of Software Engineering. IEEE Transactions on Software Engineering 25(4), 557–572 (1999)
54. Sheppard, S., Colby, A., Matacangay, K., Sullivan, W.: What is Engineering Practice? International Journal for Engineering Education 22(3), 429–438 (2006)
55. Sneed, H.N.: Software Engineering Management. Ellis Horwood, Chichester (1989)
56. Star, J.R.: On the Relationship Between Knowing and Doing in Procedural Learning. In: Proceedings of the Conference of the Learning Sciences, Ann Arbor, MI, pp. 80–86 (2000)
57. Velleman Inc., Running MicroBug (2011), http://www.vellemanusa.com
58. Wilson, B., Sherry, L., et al.: Adoption of Learning Technologies in Schools and Universities. In: Handbook on Information Technonogies for Education and Training, pp. 293–308. Springer, New York (2000)
59. Wright, P., McCarthy, J.: Empathy and Experience in HCI. In: Proceedings of the Conference on Human Factors in Computing Systems (CHI), Florence, Italy, pp. 637–646 (2008)
60. Zhao, Y., Frank, K.A.: Factors Affecting Technology Uses in Schools: An Ecological Perspective. American Educational Research Journal 40(4), 807–840 (2003)

Part III
Theoretical Refinement for Innovative Solutions

Chapter 8
FSM-Based Logic Controller Synthesis in Programmable Devices with Embedded Memory Blocks

Grzegorz Borowik, Grzegorz Łabiak, and Arkadiusz Bukowiec

Abstract. For a typical digital system, the design process consists of compilation, translation, synthesis, logic optimization, and technology mapping. Although the final result of that process is a structure built of standard cells, logic cells, macroblocks, and similar components; the characteristics of the system (the silicon area, speed, power, etc.) depend considerably on the logic model of the digital system. Therefore, the synthesis and logic optimization has a significant impact on the quality of the implementation. In this chapter, we describe methods of designing and synthesis for logic controllers in novel reprogrammable structures with embedded memory blocks. This chapter is generally based on the ideas published in [5], however, a number of issues were extended and provide detailed information about the methods and algorithms used in the problem, including [6, 13, 14, 38]. The method starts with the formal specification of a logic controller behavior. To specify the complex nature of a logic controller we have chosen statechart diagrams [21]. The main advantage of this specification is the possibility of detecting all reachable deadlocks [26]. It is particularly important in the case of safety-critical systems since any failure of such system may cause injury or death to human beings. Having graphically specified the behavior, it is subsequently converted into a mathematical model [30]. Next, the mathematical model of the statechart is transformed into an equivalent finite state machine (FSM) [29]. Thus, the logic controller in FSM form can be synthesized by applying ROM-based decomposition method [6] or architectural decomposition method [13], and finally implemented in embedded memory block equipped architectures [45, 48]. Such architectures offer ability to update

Grzegorz Borowik
Institute of Telecommunications, Warsaw University of Technology,
Nowowiejska 15/19, 00-665 Warsaw, Poland
e-mail: G.Borowik@tele.pw.edu.pl

Grzegorz Łabiak · Arkadiusz Bukowiec
Computer Engineering & Electronics Department, University of Zielona Góra,
Licealna 9, 65-417 Zielona Góra, Poland
e-mail: {G.Labiak,A.Bukowiec}@iie.uz.zgora.pl

© Springer International Publishing Switzerland 2015
R. Klempous and J. Nikodem (eds.), *Innovative Technologies in Management and Science*,
Topics in Intelligent Engineering and Informatics 10, DOI: 10.1007/978-3-319-12652-4_8

the functionality, partial reconfiguration, and low non-recurring engineering costs relative to an FPGA design.

8.1 Preliminaries

Logic controller is an electronic digital device used for automation of electrome-chanical processes, such as control of machinery on factory assembly line, lighting fixtures, traffic control systems, household appliances, and even automation of systems whose role is to maintain an ongoing interactions with their environment, i.e. controlling mechanical devices, such as a train, a plane, or ongoing processes, e.g. processes of a biochemical reactor.

Logic controller receives signals both from controlled object and optional operator, and repeatedly produces outputs to controlled object. If controller operates on binary values, as opposed to continuous values, it is called *binary control system* (binary controller). Therefore, binary controller can easily be implemented as a digital circuit. The general idea of binary control system is presented in Fig. 8.1.

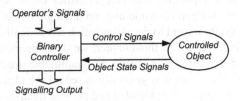

Fig. 8.1 Binary control system

An important application of binary controllers is embedded systems design [19, 20]. In particular, it is advisable for data processing systems. Thus, such system can be realized according to application-specific architecture known as a control unit with datapath (Fig. 8.2). Datapath module receives data and process them according to implemented algorithm and sends status signal to control unit. Control unit signals steers the data among the units and registers of datapath. If data are of discrete

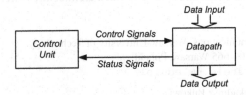

Fig. 8.2 Control unit with datapath architecture

values the whole system can be implemented in Field Programmable Gate Arrays (FPGA); the basic architecture for General-purpose processors [44].

Logic controllers can find their practical application in different areas of biotechnology. In [40] a technology of smart hand prosthesis control based on myoelectric signals is presented. The key elements of this design are arithmetic and control units. Bioreactor controller is the focus of the paper [23]. It provides an overview of current and emerging bioreactor control strategies based on unstructured dynamic models of cell growth and product formation. Nevertheless, process control plays a limited role in the biotechnology industry as compared to the petroleum and chemical industries. This demand for process modeling and control is increasing, however, due to the expiration of pharmaceutical patents and the continuing development of global competition in biochemical manufacturing. The lack of online sensors that allow real-time monitoring of the process state has been an obstruction to biochemical process control. Recent advances in biochemical measurement technology, however, have enabled the development of advanced process control systems [23].

To specify a complex nature of a controller we have chosen statechart diagrams. A statechart diagram is a state-based graphical scheme enhanced with concurrency, hierarchy and broadcasting mechanism. It may describe a complex system, but requires the system is composed of a finite number of states [21].

To synthesize statechart-base logic controller it is necessary to precisely define its behavior, however, the issue of hardware synthesis of statecharts is not solved ultimately. There are many implementation schemes depending on target technology. First, published in [17], consists of transformation of the statechart into the set of hierarchically linked FSMs traditionally implemented. In [18], a special encoding of the statechart configurations targeted at PLA structures is presented. The drawback of this method is that diagram expresses transitions between simple states only. In [36], Ramesh enhanced the coding scheme by introducing a prefix-encoding. However, the common drawbacks of the presented methods is the lack of support for history attributes and broadcast mechanism. Other implementation methods using HDL and based on ASIP are presented in [4, 24] and [10], respectively. In [16] statechart diagram is used as a graphic formalism for program specification for PLC controller, where UML language heavily support the design process.

In this chapter, we describe methods of designing and synthesis for logic controllers in novel reprogrammable structures with embedded memory blocks. It is generally based on the ideas published in [5], however, a number of issues were extended and provide detailed information about the methods and algorithms used in the problem, including [6, 13, 14, 38]. In subsection 8.2, we start with the example of a chemical reactor and its formal specification using the statechart diagram, as well as provide assumptions of hardware implementation. Having graphically specified the behavior, it is subsequently converted into a mathematical model [30]. Next, the mathematical model of the statechart is transformed into an equivalent finite state machine (FSM) [29] that is described in subsection 8.3. Finally, the logic controller in FSM form can be synthesized by applying ROM-based decomposition method [6] (see subsection 8.4.2) or architectural decomposition method [13] (see subsection 8.4.3), and finally implemented in embedded memory block equipped

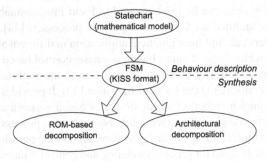

Fig. 8.3 Transformation and synthesis of logic controller

architectures [45, 48]. Such architectures offer ability to update the functionality, partial reconfiguration, and low non-recurring engineering costs relative to an
FPGA design (subsection 8.4.1). The idea of transformation and synthesis of logic
controller is provided in Fig. 8.3.

8.2   Example and Assumptions of Hardware Implementation

The statechart diagram can certainly specify reactive system behavior. As an example of practical application, the schematic diagram of a chemical reactor and appropriate statechart diagram of its logic controller are presented in Fig. 8.4 and 8.5,
respectively.

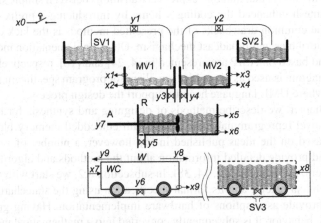

Fig. 8.4 Schematic diagram of chemical plant with wagon

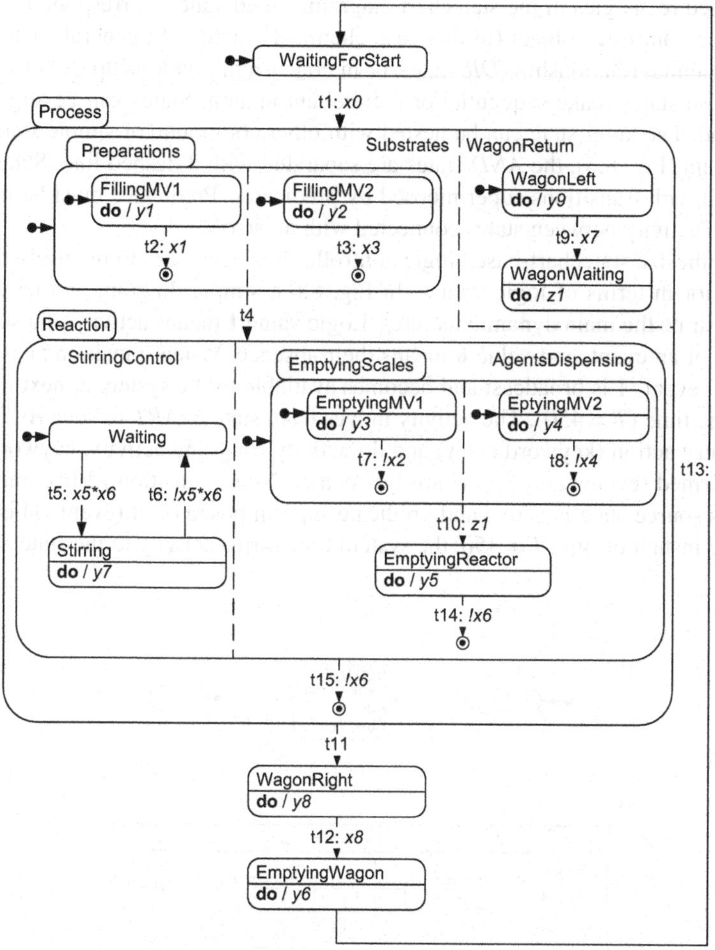

Fig. 8.5 Statechart diagram of chemical reactor controller

The working of the reactor is as follows. Initially, the reacting substances are kept in containers *SV1* and *SV2* (Fig. 8.4), and the emptied wagon waits in its initial position on the far right site (Fig. 8.5, state *WaitingForStart*). Then, the operator starts the proces with the signal *x0*. The pump *y1* and the pump *y2* make that liquid substrates from containers *SV1*, *SV2* are being measured out in scales *MV1* and *MV2*, respectively (state *Preparations*). During this, the wagon is coming back to its far left position. After the substrates are measured out, the main reaction starts (state *Reaction*). Next, scales fill main container *R* with agents (state *AgentDispensing*) and agitator *A* starts rotating (state *StirringControl*). After filling up the main container, the product of the reactor is poured to the wagon (state *EmptyingReactor*). Then, the wagon goes to empty (states *WagonRight* and *EmptyingWagon*).

Rounded rectangles in the statechart diagram, called states, correspond to activities in the controlled object (in this case chemical reactor). In general, states can be in sequential relationship (*OR* state), or in concurrent relationship (*AND* states). Then, these states make sequential or parallel automaton. States can be simple or compound. The latter state can be nested with other compound or simple states. In the diagram (Fig. 8.5), the *AND* states are separated with a dashed line. States are connected with transitions superimposed by predicates. Predicates must be met to transform activity between states connected with an arrow.

To synthesize statechart-based logic controller it is necessary to precisely define its behavior in terms of logic values. In Fig. 8.6 a simple diagram and its waveform illustrate the main dynamic features. Logic value **1** means activity of a state or presence of an event, and value **0** means their absence. When transition $t1$ is fired ($T = 350$) event $t1$ is broadcast and becomes available to the system at next instant of discrete time ($T = 450$). The activity moves from state *START* to state *ACTION*, where entry action (keyword *entry*) and do-activity (ongoing activity, keyword *do*) are performed (events *entr* and *d* are broadcast). Now, transition $t2$ becomes enabled. Its source state is active and predicate superimposed on it (event $t1$) is met. So, at the instant of time $T = 450$, the system transforms activity to the state *STOP*,

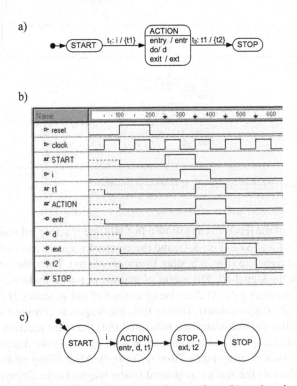

Fig. 8.6 Simple diagram with main features (a), its waveform (b), and equivalent FSM (c)

performs exit action (keyword *exit*, event *ext*) and triggers event $t2$, which do not affect any other transition. The step is finished.

Summarizing, dynamic characteristics of hardware implementation are as follows:

- system is synchronous,
- system reacts to the set of available events through transition executions,
- generated events are accessible to the system during next tick of the clock.

Basing on the assumptions, hardware implementation involves mapping syntax structure of statechart into hardware elements, namely, every state and action (transition actions, *entry* and *exit* actions) are assigned flip-flops and every event is a signal. Both flip-flops and signals have their functions (flip-flops have excitation functions) which are created based on rule of transitions firing and on presented hardware assumption [27, 28].

8.3   Transformation to FSM Model

The transformation of statechart diagram into FSM model [22] involves building equivalent FSM Mealy automaton using statechart elements which for external observer behaves just the way statechart does. Final Mealy automaton is described in KISS format [42]. The main notion the transformation revolves around is a global state of the statechart [29, 30]. The general idea is to create an equivalent target Mealy automaton whose states corresponds to the global states of a statechart. The global state of a statechart is defined as a superposition of local activities, namely, local states, transition events, *entry* and *exit* actions. In hardware implementation these activities are implemented by means of flip-flops and their excitation functions (see subsection 8.2). The whole process is explained on the example diagram in Fig. 8.7.

The diagram in Fig. 8.7 features both concurrency and hierarchy. It consists of five simple states $\{s_1, s_3, s_4, s_5, s_6\}$, one compound state s_2, and two *do* actions $\{x, y\}$ assigned to states, s_4 and s_6, respectively. The diagram has three input signals $\{a, b, c\}$ and two output signals $\{x, y\}$. There is no transition action nor *entry* and *exit* actions.

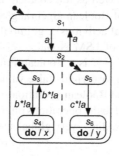

Fig. 8.7 Simple concurrent and hierarchical diagram

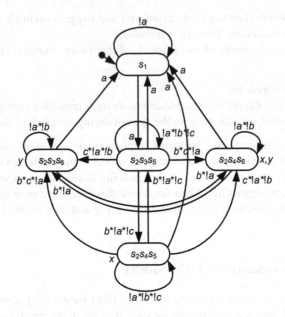

Fig. 8.8 Equivalent FSM Moore automaton of diagram in Fig. 8.7

Having defined a global state of a statechart as a vector of local activities, it is possible to construct hypothetical equivalent Moore type automaton. Fig. 8.8 presents such an automaton. Names of the states correspond to local activities of the original diagram, e.g. state $s_2s_4s_6$ corresponds to the activity of the statechart states $\{s_2, s_4, s_6\}$, respectively, and broadcast events $\{x, y\}$. Transition labels are Boolean expressions which must be met to fire transitions. It is necessary to emphasis that both models statechart and FSM are deterministic and their transition predicates must be orthogonal. In general, due to concurrency and hierarchy, equivalent FSMs have more states and transitions, which is a big drawback of the transformation.

The vector of local activities with their excitation functions allows us to symbolically express a global state as well as the characteristic function for every global state of a statechart. In terms of Boolean algebra, global state is defined as a product of signals representing local activities, e.g. initial global state of the statechart which corresponds to state s_1 of FSM is $G_0 = s_1\bar{s}_2\bar{s}_3\bar{s}_4\bar{s}_5\bar{s}_6$. Characteristic function χ_G of a set of elements $G \subseteq U$ is a Boolean function which yields true value ($\chi_G(g) = 1$) when $g \in G$, otherwise yields false value. It means that a characteristic function represents a set, and Boolean operations on the characteristic functions (e.g. $+$, $*$, negation) correspond to the operations on sets which they precisely represent (e.g. $\cup$, $\cap$, complement). Coding every global state of a statechart, just like G_0, and summing them we obtain characteristic function representing every global state of a statechart. For the diagram in Fig. 8.7 the characteristic function is as follows:

$$\chi_{[G_0]} = s_1\bar{s}_2\bar{s}_3\bar{s}_4\bar{s}_5s_6 + \bar{s}_1s_2s_3\bar{s}_4\bar{s}_5s_6 + \bar{s}_1s_2s_3\bar{s}_4s_5\bar{s}_6 +$$
$$+ \; \bar{s}_1s_2\bar{s}_3s_4\bar{s}_5s_6 + \bar{s}_1s_2\bar{s}_3s_4s_5\bar{s}_6 \qquad (8.1)$$

Computation of the characteristic function is carried out according to the algorithm 8.1 performing symbolic traversal.

Listing 8.1 Symbolic traversal of state space

```
symb_trav_of_Statechart (Z, init_state) {
    χ[G0] = curr_states = init_state;
3   while (curr_states != 0) {
        next_states = Image_computation(Z, curr_states);
5       curr_states = next_states * χ[G0];
        χ[G0] = curr_states + χ[G0];
7   }
}
```

Sets of global states are represented by their characteristic functions (*put in italic*). The operations on the sets are represented by Boolean operations on corresponding characteristic functions. The algorithm starts with *init_state* (e.g. G_0) and in breath-first manner generates the set of all reachable global next states in one formal step (line no. 4). In line no. 5, only new global states are calculated (*curr_states*), and in line no. 6, new states are added to the so far generated global states ($\chi_{[G_0]}$). The algorithm stops when there is no new *curr_states* (line no. 3).

The formal step from line no. 5 is performed by the *Image_computation* procedure. For the *curr_states* it generates their image in the transformation with the functional vector (δ_S) of excitation functions of local activities in the statechart diagram. Hence, it is possible, in one formal step to compute the set of all reachable global next states for the set of global states (*curr_states*). The procedure is as follows:

$$next\_states = \exists_s\exists_x(curr\_states * \prod_{i=1}^{n}[s'_i \odot (curr\_states * \delta_{S_i}(s,x))]), \quad (8.2)$$

$$next\_states = next\_states\langle s' \leftarrow s\rangle, \qquad (8.3)$$

where s, s', x denote the present state, next state and input signals, respectively; $\exists_s$ and $\exists_x$ represent the existential quantification of the present state and signal variables; n is a number of state variables; $\odot$ and $*$ represent logic operators XNOR and AND, respectively; equation 8.3 means swapping variables in expression.

Characteristic function does not contain information on transitions. This information can be obtained by the calculation of the transition relation under input (χ_{TRI}) using the following equation:

$$\chi_{TRI}(s',x,s) = \prod_{i=1}^{n}[s'_i \odot \delta_{S_i}(x,s)], \qquad (8.4)$$

where the symbol $\odot$ represents the logic XNOR operator and n is the number of state variables. The relation $\chi_{TRI}(s',x,s) = 1$ implies that in state s, there exists a transition to state s' on input x.

The transformation of Moore automaton (Fig. 8.8) into Mealy automaton (Fig. 8.9) is very simple [22], but we must add one extra state (*start*) to include output for initial state. The two characteristic functions ($\chi_{[G_0]}$ and χ_{TRI}) provide enough information to generate equivalent FSM in KISS format. The algorithm 8.2 starts with characteristic function of global states space $\chi_{[G_0]}$ and with characteristic function χ_{TRI}. The transition regarding input signals is represented by the product t, which is a relation between current (G_i) and next (G'_j) states, represented as conjunction formulae (line no. 4). For every pair of states: current state and next state (G_i, G'_j), it is being checked whether there is a transition between them (line no. 5). In line no. 6 state variables (s, s') are removed from transition product t, hence t_x represents the only part of the expression which solely depends on input variables x. Thus, current and next states are put into 4-tuple KISS line (lines 7 and 8) and sent into KISS file (line no. 21). Between lines 15 and 20 the input vector is being computed and dependency on input variables is being checked for each minterm in t_x expression. It is formally computed according the following formula:

$$\frac{\partial f}{\partial x_i} = f_{x_i} \oplus f_{\bar{x}_i}, \tag{8.5}$$

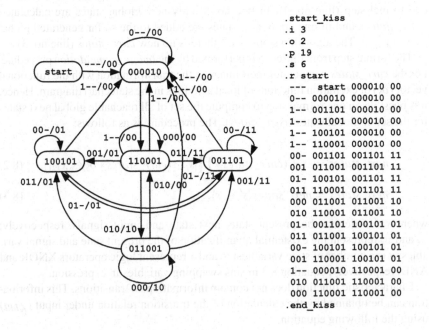

```
.start_kiss
.i 3
.o 2
.p 19
.s 6
.r start
---   start  000010 00
0--  000010 000010 00
1--  001101 000010 00
1--  011001 000010 00
1--  100101 000010 00
1--  110001 000010 00
00-  001101 001101 11
001  011001 001101 11
01-  100101 001101 11
011  110001 001101 11
000  011001 011001 10
010  110001 011001 10
01-  001101 100101 01
011  011001 100101 01
00-  100101 100101 01
001  110001 100101 01
1--  000010 110001 00
010  011001 110001 00
000  110001 110001 00
.end_kiss
```

Fig. 8.9 KISS of the FSM in Fig. 8.8

Listing 8.2 Transitions generation algorithm

```
Transitions_Generation(Z, χ[G₀] ,χTRI) {
2  for each global state Gᵢ χ[G₀] (Gᵢ) = 1;
     for each global state G'ⱼ χ[G₀] (G'ⱼ) = 1; {
4      t = χTRI(s',x,s) * Gᵢ * G'ⱼ ;
       if t = 0 then continue ;
6      tₓ = ∃s'∃st
       current−st = Gᵢ ;
8      next−st = G'ⱼ⟨s' ← s⟩ ;
       for each output yᵢ ∈ Y {
10       if G'ⱼ⟨s' ← s⟩ * λMᵢ ≠ 0 then out[i] = 1
         else out[i] = 0;
12     }
       for each minterm mᵢ in tₓ {
14       for each input xⱼ ∈ X {
           if ∂mᵢ/∂xⱼ ≠ 0 then {  // deps on xⱼ
16           if mᵢ*xⱼ ≠ 0 then in[j] = 1
             else in[j] = 0;
18         }
           else in[j] = −;
20       }
         KISS << <in , current−st , next−st , out>
22     }
     }
24 }
```

where f_{x_i}, $f_{\bar{x}_i}$ are positive and negative algebraic cofactor, respectively, and the symbol $\oplus$ represents XOR operator. In lines 16, 17 and 19 we determine the impact of signal x_j on the transition. Subsequently, in lines from 9 to 11, we compute the output vector, where y_i is an i-th output variable, and λ_i is its signal function. Although this is not presented in the algorithm, it is enough to execute these four lines only once per the next state s'.

Figure 8.9 presents final Mealy automaton and its KISS file of the diagram from Fig. 8.7. Binary labels both, states and transitions, correspond to signals and states activities from statechart diagram according to the following order: input $[a,b,c]$, output $[x,y]$, state $[s_3,s_5,s_4,s_6,s_1,s_2]$. A single line in KISS file, e.g. `1-- 000010 110001 00`, defines transition from state s_1 (`000010`) to state $s_2 s_3 s_5$ (`110001`) under the input a without active outputs.

The transformation has been successfully implemented in academic *HiCoS* system [47] and all Boolean transformations have been performed by means of Binary Decision Diagrams [33, 35].

8.4 Synthesis

In modern logic synthesis, regardless of the implementation technology (programmable devices, Gate Array or Standard Cell structures), the problem of finite state machine synthesis (in particular – the problem of internal state encoding) is an issue of significant practical importance.

Many methods for structural synthesis of FSMs have been reported in the literature. Their diversity is a consequence of different assumptions taken to simplify calculations, as well as different types of intended target components. Thus, different methods of FSM synthesis have been developed for PLA structures [32, 34, 41], for ROM memories [6, 37, 38], and for PLD modules [15].

A distinctive feature of traditional methods of FSM synthesis is the application of logical minimization before the process of state encoding. This minimization is possible when the inputs and outputs of the combinational part of the sequential circuit is represented with multi-valued symbolic variables. Unfortunately, such methods are limited to two-level structures. For other implementation styles different methods are needed. The research in this area goes into two directions: one concerns the implementation with multilevel gate structures, while the other embraces implementations with cellular FPGA and CPLD structures.

In the first case, like for two-level structures, the starting point of the synthesis process is a structure in which the combinational circuit is connected to the inputs of a register operating as state memory (Fig. 8.10a), whereas in the other case, the combinational circuit is connected to the outputs of such a register (Fig. 8.10b).

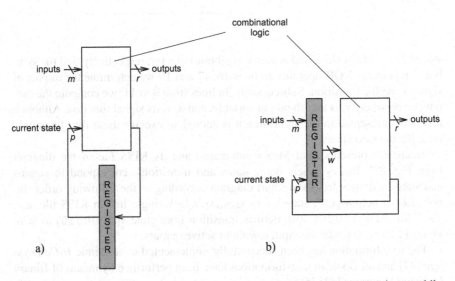

Fig. 8.10 Two models of a sequential circuit: classical (a), with microprogramming capability (b)

Until recently, mainly the first model (Fig. 8.10a) was used in synthesis of sequential machines. The optimization of the selection of state encoding was done for two-level or multilevel gate structures and was aimed at the reduction of hardware resources (silicon area).

The second model (Fig. 8.10b) was used in microprogrammed control circuits, with the combinational circuit implemented with ROM memory [1]. In the microprogrammed version of the sequential circuit, the fixed ROM memory was a separate element – separated from the rest of the circuit. The advantage of this structure was an ability to program the microcode memory, which was the only possible way to reconfigure the circuit at that time. These advantages made the capacity of the memory to be a non-critical factor, although the reduction of this capacity was a common optimization criterion.

Microprogrammed control has been a very popular alternative implementation technique for control units. However, as systems have become more complex and new programmable technologies have appeared, the concept of classical microprogramming has become less attractive for control unit implementations. But the main idea of Microprogrammed Control Units, i.e. implementation of combinational part of the sequential circuit with a ROM, has gained new motivation after the appearance of programmable logic devices [3]. In particular, the growing interest in ROM-based synthesis of finite state machines has been caused by the inclusion of Embedded Memory Blocks in modern FPGAs.

8.4.1 Modern Technologies of Controller Manufacture

Field-programmable devices are very often used for the implementation of logic controllers. Since these devices can be programmed by the user during the design process, they are a good platform for dedicated control algorithms. There are many different types of such devices – from simple Programmable Logic Devices (PLDs) through Complex PLDs (CPLDs) to advanced Field Programmable Gate Arrays (FPGAs) [25].

The research presented in this chapter is oriented towards FPGA devices with Embedded Memory Blocks.

FPGAs are built with a matrix of small configurable logic blocks (CLBs), which are connected using internal programmable interconnections and surrounded by programmable input/output blocks (IOBs) for communication with the environment (Fig. 8.11) [25]. An FPGA contains from several to tens of thousands of CLBs. Each logic block is built of look-up tables (LUTs), D type flip-flops, and some additional control logic. One LUT has typically four inputs, however up to six or more [39], and can implement any Boolean function of this number of variables, i.e. working of four-input LUT can be perceived as 16×1 ROM.

The new FPGAs have also memory blocks [45, 48]. They have different names depending on vendor, for e.g. Embedded Memory Block (EMB) in Altera devices or

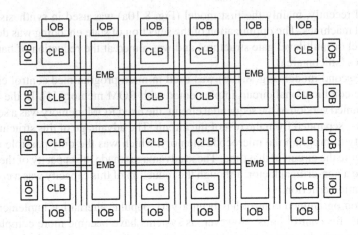

Fig. 8.11 Structure of an FPGA device

Block RAM in Xilinx devices. These blocks are placed in columns typically in outer areas of device (Fig. 8.11). These memories are from 512 bits up to 36 Kbits in size. The most popular size of the memory block of cheaper FPGAs is 4 Kbits. The very big advantage of such block is existence of feature that allows them to be set to one of several modes of data width (Tab. 8.1). They can also work in one of modes, like single-port RAM, a dual-port RAM or ROM. When an embedded memory block works in the ROM mode, it is initiated with the content during the programming process of an FPGA device. In this mode, it can be used for the implementation of combinational functions.

Table 8.1 Typical configuration modes of 4-Kbit embedded memory block

Mode	Number of words	Width of the word [bits]
4K×1	4096	1
2K×2	2048	2
1K×4	1024	4
512×8	512	8
256×16	256	16

8.4.2   Functional Decomposition and ROM-Based Synthesis

One of the main goals of the synthesis is not only technological implementation of logic controller but also optimization of hardware resources consumption. It is particularly important when the design is intended for novel programmable structure containing LUT-based cells and embedded memory blocks. The other factor of vital importance is Boolean minimization strategy. Authors' proposition is to apply the idea of functional decomposition, i.e. a structure with address modifier (Fig. 8.12b), which is best suited for the FSMs in KISS format from previous section [6, 37, 38].

A limited size of embedded memory blocks available in FPGAs is the main reason behind the application of this structure. The implementation of an FSM shown in Fig. 8.12b can be seen as a serial decomposition of the memory block included in the structure of Fig. 8.12a into two parts: an address modifier and a memory block of smaller capacity than required for the realization of the structure in Fig. 8.12a. In the considered FSM implementation both parts are implemented in embedded memory blocks which are configured as ROM memory, with its content determined at the time of the programming. The address modifier (first block) is used here to reduce the number of memory address lines of the second block.

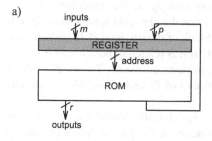

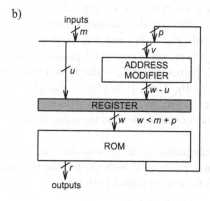

Fig. 8.12 FSM implementation: using ROM memory (a), with the addition of an address modifier (b)

As a result, sequential circuits requiring large-capacity ROM memories (and thus not implementable in the architecture of Fig. 8.12a) can be implemented using memory blocks with a smaller number of inputs. Then, the size of the required memory [6] is equal

$$M = 2^v \cdot (w - u) + 2^w \cdot (r + p).$$ (8.6)

For explaining the work of functional decomposition with address modifier partition description and partition algebra are applied [22]. They describe logic dependencies in considered FSM.

Let $\mathbb{T}$ be an isomorphic function between the domain D_δ of the transition function and the set $T = 1, \ldots, t$, where $t = |D_\delta|$. Set T represents the ROM cells needed to store the next state pair $\delta(v, s)$ for each pair (v, s). A *characteristic partition P_c* of the FSM is defined in the following way: each block B_{P_c} of the characteristic partition includes these elements from the set T which correspond to these pairs (v, s) from the domain D_δ which the transition function $\delta(v, s) = s'$ maps onto the same next state s'.

A partition P on the set T is related to a partition π on the states set S if for any inputs v_a, v_b the condition that s_i, s_j belong to one block of the partition π implies that the elements from T corresponding to pairs (v_a, s_i) and (v_b, s_j) belong to one block of the partition P. A partition P on set T is related to a partition θ on the input symbols set V if for any state s_a, s_b the condition that v_i, v_j belong to one block of the partition implies that the elements from T corresponding to pairs (v_i, s_a) and (v_j, s_b) belong to one block of the partition P. In particular, a partition P on set T is related to the set $\{\pi, \theta\}$ if it is related to both π and θ.

Let P_a and P_b be partitions on the set T, and $P_a \geq P_b$. Then a partition $P_a|P_b$, whose elements are blocks of P_b and whose blocks are those of P_a, is a *quotient partition* of P_a over P_b.

For a partition $P_a \geq P_b$ let $P_a|P_b$ denote the quotient partition and let $\varepsilon(P_a|P_b)$ be the number of elements in the largest block of partition $P_a|P_b$. Let $e(P_a|P_b)$ be the smallest integer equal to or greater than $\log_2(\varepsilon(P_a|P_b))$ (i.e., $e(P_a|P_b) = \lceil \log_2(\varepsilon(P_a|P_b)) \rceil$). Then, the notion of *r-admissibility* of the two-block partitions' set $\{P_1, \ldots, P_k\}$ on S in relation to the partition P on S, is defined as $r = k + e(\sigma|\rho)$, where σ is the product of $\{P_1, \ldots, P_k\}$ and ρ is the product of σ and P.

Let $\Pi = \{\pi_1, \ldots, \pi_p\}$ is the set of two-block partitions on S and $\Theta = \{\theta_1, \ldots, \theta_m\}$ is the set of two-block partitions on V, while P_k is a partition on the set T which is related to either π_i or θ_j. Then, $\mathbb{p} = \{P_1, \ldots, P_{m+p}\}$ is the set of all partitions related to partitions $\{\pi_1, \ldots, \pi_p, \theta_1, \ldots, \theta_m\}$. Partitions in Π correspond to the state variables and partitions in Θ correspond to the input variables.

Fact. To achieve unambiguous encoding of address variables and at the same time maintain the consistency relation $\mathbb{T}$ with the transition function, two-block partitions $\mathbb{P} = \{P_1, \ldots, P_w\}$ have to be found, such that:

$$P_1 \cdot P_2 \cdot \ldots \cdot P_w \leq P_c.$$ (8.7)

This is a necessary and sufficient condition for $\{P_1, \ldots, P_w\}$ to determine the address variables. This is because each memory cell is associated with a single block of P_c, i.e., with those elements from T which map the corresponding (v, s) pairs onto the same next state.

Although some of the partitions for the $\mathbb{P}$ set can be selected from the p set, the selection is made in such a way that the simplest addressing unit (address modifier) is produced. Such a selection is possible thanks to the method [9], based on the notion of r-admissibility.

Assume that u partitions $\{\pi_1, \ldots, \pi_l\}$ and $\{\theta_1, \ldots, \theta_{u-l}\}$ were chosen. These partitions correspond to the address lines driven by a single variable, either a state variable q or an external variable x. The result is the state and input symbol partial encoding; i.e.,

$$a_1 = q_1, \ldots, a_l = q_l, a_{l+1} = \theta_1, \ldots, a_u = \theta_{u-l}.$$

This encoding of state variables is possible thanks to the method of construction and coloring weighted graphs.

Then, inequality (8.7) can be written as:

$$P_{i_1} \cdot P_{i_2} \cdot \ldots \cdot P_{i_u} \cdot P_{i_{u+1}} \cdot \ldots \cdot P_{i_w} \le P_c, \tag{8.8}$$

where $P_U = P_{i_1} \cdot P_{i_2} \cdot \ldots \cdot P_{i_u}$ is related to the partitions $\{\pi_1, \pi_2, \ldots, \pi_l, \theta_1, \theta_2, \ldots, \theta_{u-l}\}$.

The encoding of the part of the state variables remaining after the partial encoding (input variables, in general) can be obtained from the following rules:

$$\pi_1 \cdot \pi_2 \cdot \ldots \cdot \pi_l \cdot \pi = \pi(\mathbf{0}),$$
$$\theta_1 \cdot \theta_2 \cdot \ldots \cdot \theta_{u-l} \cdot \theta = \theta(\mathbf{0}),$$

where π and θ represent partitions corresponding to these remaining state variables. $\pi(\mathbf{0})$ as well as $\theta(\mathbf{0})$ are partitions whose blocks are equal to their elements.

Since the design process may be considered as a decomposition of the memory block into two blocks: a combinational address modifier and a smaller memory block, we need to find function G which will determine the second part of the memory address bits.

Inequality (8.8) can be transformed into:

$$P_U \cdot P_G \le P_c. \tag{8.9}$$

Now, a partition P_G has to be constructed, such that:

$$P_G \ge P_V, \tag{8.10}$$

where $P_G = P_{i_{u+1}} \cdot \ldots \cdot P_{i_w}$ and P_V is related to the partition set $\{\pi, \theta\}$. Let us assume that input variables are encoded.

Theorem. Partition P_V can be constructed in the following way:

$$P_V = P_S \cdot P_{V_\theta}, \tag{8.11}$$

where P_S is the partition related to $\pi(0)$ on the set of states S, and P_{V_θ} is the partition related to θ.

Proof. Let us assume that $P_V = P_{V_\pi} \cdot P_{V_\theta}$, where P_{V_π} is related to π, and P_{V_θ} is related to θ. Since P_U and P_V satisfy $P_U \cdot P_V \le P_c$, we have $P_U \cdot P_{V_\pi} \cdot P_{V_\theta} \le P_c$. As a result, $P_U \cdot P_S \cdot P_{V_\theta} \le P_c$.

Let $\langle V, R, E, P \rangle$ be a quadruple where: V – set of elements, R – an equivalence relation on the set V, E – set of pairs in relation P on the set V, P – two-element relation. A triple $M(V|R, E, P)$ is a *multi-graph*, where $V|R$ – is an equivalence class for an equivalence relation on the set V. Since there exists an isomorphism $V|R \leftrightarrow V'$, we can construct a natural mapping from the multi-graph $M(V|R, E, P)$ to the graph $G(V', E', P)$. This mapping $\psi: M \to G$ allows for calculation of a chromatic number $\chi(G) = \chi(M)$.

Inequality (8.9) allows us to construct a quotient partition $P_U|P_c$. Then, the triple $\langle P_V, E_1, P_1 \rangle$, where: P_V is a partition given by equation (8.11), P_1 is a relation which represents incompatibilities in quotient partition $P_U|P_c$ on the set T (relation of incompatibility in quotient partition $P_U|P_c$ is a relation among all elements in each block of the partition separately) and E_1 is the set of pairs in the relation P_1; is a multi-graph $M_1(P_V, E_1, P_1)$.

After mapping $\psi_1: M_1 \to G_1$ we calculate a chromatic number $\chi(G_1)$ which is equal to $\chi(M_1)$. The coloring of the graph G_1 determines the P_G partition. The value of

$$\mu = |U| + \lceil \log_2(\chi(M_1)) \rceil \tag{8.12}$$

determines memory size required.

In case of $\mu > w$, a new partition P_V' has to be constructed. Then, P_V has to be multiplied by appropriately chosen two-block partitions related to those which are generated by input variables from the set U. In that case, we obtain a non-disjoint decomposition [8].

In the next step we calculate the remaining state variables. The triple $\langle P_S, E_2, P_2 \rangle$, where: P_S is the partition related to $\pi(0)$ on the states set S, P_2 is a relation which represents incompatibilities in quotient partition $P_{V_\theta}|P_G$ and E_2 is the set of pairs in the relation P_2; is a multi-graph $M_2(P_S, E_2, P_2)$.

Similarly to the case discussed above, by coloring an image graph G_2 for the multi-graph M_2, we obtain a new partition on the set S. We encode this partition with the minimal binary code. Value $\lceil \log_2(\chi(M_2)) \rceil$ determines the number of bits needed to encode the remaining state variables and value

$$\nu = |V_\theta| + \lceil \log_2(\chi(M_2)) \rceil \tag{8.13}$$

determines the number of inputs to address modifier.

The issue of finite state machine functional decomposition for FPGA structures with embedded memory blocks has been successfully implemented in university software FSM*dec* [46].

Applying the software to the chemical reactor from Fig. 8.4 transformed to KISS format (Fig. 8.13) we have obtained the content of the address modifier and ROM, as well as state code assignment (Fig. 8.14).

```
.i 10
.o 9
.p 263
.s 33
.r strst
(...)
---1---0-- s7   s6 000001000
-------0-1 s8   s6 000001000
---1---0-1 s9   s6 000001000
-------1-- s6  s10 000000000
---1---1-- s7  s10 000000000
-------1-1 s8  s10 000000000
---------1 s11 s10 000000000
(...)
```

Fig. 8.13 Chemical reactor from Fig. 8.4 transformed to KISS format

a)
```
.i 13
.o 2
.p 57
.type fr
(...)
1-------01111 00
-----0-000111 00
01----0--0000 00
01------10001 00
-0----0-100-- 01
-0------100-1 01
-0------1001- 01
(...)
```

b)
```
.i 9
.o 19
.p 87
.type fr
(...)
-00001101 0001100101000011000
-00110101 0001100101000011000
--0000110 0011010011000000000
--0001001 0011010011000000000
-1000-110 0011010011000000000
-10010101 0011010011000000000
-10110111 0011010011000000000
(...)
```

c)
```
(...)
s7  := 0001100101
s8  := 0000100110
s9  := 0110100111
s10 := 0011010011
s11 := 0001001000
s12 := 0010101000
s13 := 0101001001
(...)
```

Fig. 8.14 Content of address modifier (a), ROM (b), and state code assignment (c); after functional decomposition of chemical reactor

In Fig. 8.15 we provide results for FSM from Fig. 8.9.

a)   b)   c)

```
.i 5                        .i 4                        start   := 0000
.o 2                        .o 6                        000010  := 0001
.p 14                       .p 11                       001101  := 1010
.type fr                    .type fr                    011001  := 1011
--000 00                    11-- 000100                 100101  := 0100
0-0-0 00                    1-1- 000100                 110001  := 1101
0101- 00                    -000 000100
11101 00                    000- 000100
--001 01                    1001 110100
000-1 01                    0110 110100
10-01 01                    0100 101011
0-100 10                    0011 101011
0010- 10                    0101 101110
10011 10                    0111 010001
1-010 11                    0010 010001
1-100 11                    .e
1101- 11
01101 11
.e
```

Fig. 8.15 Content of address modifier (a), ROM (b), state code reassignment (c)

8.4.3   Synthesis Based on Architectural Decomposition

Other method which cut down the hardware implementation resources for FSMs is architectural decomposition [1, 3]. In this method, the FSM circuit is implemented in a double- or multi-level structure. The circuit of single-level is a combinational circuit that implements Boolean functions of the decomposed FSM. In comparison to the single-level circuit the gain on this circuit is that it implements less Boolean functions and typically requires less look-up tables for its implementation. The second-level circuit works as a decoder. Functions describing its behavior have a regular structure. It means that it can be implemented into new FPGA devices with the use of embedded memory blocks. Overall, such a circuit requires less logic elements but additional memory resources, although memories in FPGAs are very often not used for any other purpose.

Special methods of encoding [3] and modifications of a logic circuit structure are applied. Presented method is based on multiple encoding of internal states [12] and microinstructions [11]. The encodings are performed independently to each other and they can be applied both together [13] or separately [11, 12]. In the former solution the current state is used as partitioning set. In this case, the code of a microinstruction $K(Y_t)$ is represented by concatenation of the multiple code of the microinstruction $K_s(Y_t)$ and the code of the current state $K(s)$:

$$K(Y_t) = K_s(Y_t) * K(s). \tag{8.14}$$

The code of the internal state $K(s')$ is represented by concatenation of the multiple code of the internal state $K_s(s')$ and the code of the current state $K(s)$:

$$K(s') = K_s(s') * K(s). \tag{8.15}$$

A digital circuit of a FSM with such encodings can be implemented as a double-level structure (Fig. 8.16) called PAY_0 [13]. In this structure, the combinational circuit implements logic functions that encode microinstructions:

$$\lambda_1 : X \times Q \to \Psi, \tag{8.16}$$

and internal states:

$$\delta_1 : X \times Q \to T, \tag{8.17}$$

where Q is the set of variables storing the code of current state, and $|Q| = p = \lceil \log_2 |S| \rceil$. T is the set of variables storing the code of internal state, and $|T| = p_1$. Ψ is the set of variables storing the code of microinstruction, and $|\Psi| = r_1$. The instruction decoder implements a function of a decoder of microinstructions:

$$\lambda_2 : \Psi \times Q \to Y \tag{8.18}$$

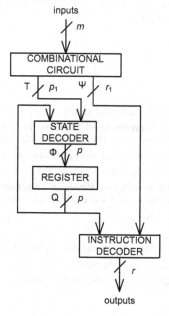

Fig. 8.16 The structural diagram of PAY_0 Mealy FSM

and functions λ_1 and λ_2 yield the function λ of mathematical transformations. The state decoder decodes internal states and generates excitation function:

$$\delta_2 : T \times Q \to \Phi \tag{8.19}$$

and, in similar way, functions δ_1 and δ_2 yield the function δ of mathematical transformations. These mathematical transformations proof that application of architectural decomposition do not change the behavior of FSM. The register is build with D-type flip-flops.

The starting point for synthesis process with architectural decomposition is the KISS file obtained from transformation of the statechart, and it consists of the following steps:

- multiple encoding of microinstructions and internal states,
- formation of the transformed table,
- formation of the system of Boolean functions,
- formation of the instruction decoder table,
- formation of the state decoder table,
- implementation of the logic circuit of the FSM.

The multiple encoding of microinstructions and internal states is based on binary encoding of microinstructions and internal states in each subset.

In the case of chemical reactor example (Fig. 8.4, Fig. 8.13), there are 33 states. It means that there are 33 subsets of microinstructions and 33 subsets of internal states. There are maximum 16 microinstructions in one subset. It means that for encoding only 4 bits are required. For example, for the subset based on the state $s8$, the following encoding is obtained:

$K_{s8}(000001000) = 0000,$
$K_{s8}(000000000) = 0001,$
$K_{s8}(000001100) = 0010,$
$K_{s8}(000000100) = 0011.$

Similar situation is obtained for internal states.

The formation of the transformed table is the base for forming system of Boolean functions. It is created from the original table (described in the KISS file) by replacing the state, the microinstruction and the internal state with their codes. Part of transformed KISS table is shown in Fig. 8.17.

The formation of the system of Boolean functions is the base for obtaining Boolean functions. These systems are obtained from the transformed table, in a classical way [1]. For example:

$$\psi_1 = x_1 x_2 x_3 x_4 \bar{x}_9 \bar{Q}_1 Q_2 \bar{Q}_3 Q_4 Q_5 \bar{Q}_6 + x_1 \bar{x}_2 \bar{x}_3 x_4 \bar{x}_9 \bar{Q}_1 Q_2 \bar{Q}_3 Q_4 Q_5 \bar{Q}_6$$
$$+ x_1 x_2 \bar{x}_3 x_4 \bar{x}_9 \bar{Q}_1 Q_2 \bar{Q}_3 Q_4 Q_5 \bar{Q}_6 + x_1 x_2 x_3 x_4 \bar{x}_9 \bar{Q}_1 Q_2 \bar{Q}_3 Q_4 Q_5 \bar{Q}_6$$
$$+ \bar{x}_1 x_2 x_3 x_4 \bar{x}_9 \bar{Q}_1 Q_2 \bar{Q}_3 \bar{Q}_4 Q_5 \bar{Q}_6 + \bar{x}_1 \bar{x}_2 \bar{x}_3 x_4 \bar{x}_9 \bar{Q}_1 Q_2 \bar{Q}_3 Q_4 Q_5 \bar{Q}_6$$
$$+ \ldots$$

```
.i 10
.o 9
.p 263
.s 33
.r strst
(...)
 ---1---0-- 000110 0000 0000
 -------0-1 000111 0001 0000
 ---1---0-1 001000 0000 0000
 -------1-- 000101 0001 0001
 ---1---1-- 001001 0001 0001
 -------1-1 000111 0001 0001
 ---------1 001010 0000 0001
(...)
```

Fig. 8.17 Example of transformed KISS description

The formation of the instruction decoder table. This step forms the table that describes the behavior of instruction decoder. This table has four columns:

- binary code of the current state;
- binary code of the microinstruction (from adequate subset);
- binary representation of the microinstruction (from adequate subset);
- number of the line.

Table 8.2 shows part of such table for our example.

Table 8.2 The part of the instruction decoder table

$K(s)$	$K_s(Y_t)$	Y_t	h
000000	0000	000000000	1
000001	0000	010000000	2
000001	0001	000000000	3
000010	0000	010000000	4
000010	0001	000100000	5

The formation of the state decoder table. This step forms the table that describe behavior of the state decoder. This table has four columns:

- binary code of the current state;
- binary code of the internal state (from adequate subset);
- binary representation of excitation functions that switches the memory of the FSM;
- number of the line.

Table 8.3 The part of the state decoder table

$K(s)$	$K_s(s')$	D	h
000000	0000	000011	1
000001	0000	000001	2
000001	0001	000011	3
000010	0000	000001	4
000010	0001	000010	5

Table 8.3 shows part of such table for the presented example.

The implementation of the logic circuit of the FSM. The combinational circuit and the register are implemented with CLBs – LUTs and D-type flip-flops. The instruction decoder is implemented using embedded memory blocks with $2^{(p+r_1)}$ words of r bits. The content of the memory is described by the instruction decoder table, where concatenation of the binary code of the current state and the binary code of a microinstruction (Fig. 8.14) is the memory address, and the binary representation of a microinstruction is the value of the memory word. The state decoder is also implemented in an embedded memory block with $2^{(p+p_1)}$ words of p bits. The content of the memory is described by the state decoder table, where concatenation of the binary code of the current state and the binary code of the internal state (Fig. 8.15) is the memory address, and the binary representation of excitation functions is the value of the memory word. Any (*don't care*) value can be assigned for addresses missing in both tables, since such concatenations of codes for both memories are never used. Both decoders are implemented by memory blocks of an FPGA device working in ROM mode. Schematic diagram for an FPGA technology of multi-level structures is presented in Fig. 8.18.

This diagram is based on Xilinx Spartan and Virtex FPGAs but they can be easy adopted to other FPGAs vendors, like Altera Cyclone and Stratix, since all logic elements, especially memory blocks and their connections, are very similar.

It should be mentioned, that memory blocks in popular FPGAs are synchronous [45, 48]. The clock signal for memory blocks is the same as for the register but memory blocks are trigged by opposite edge (in this case falling edge). It cause that data are ready to read after one cycle and there is no need to wait one clock cycle more until data are stable. It is especially important when an internal state is encoded. It also means that memory blocks can work as an output register in case when microoperations are encoded. Such registers are needed in each digital system with Mealy's outputs to stabilize its operation [2, 25]. Other input signals of memory blocks are connected to logic 1 or logic 0, according to specification of Xilinx Block RAM [48], to satisfy read-only mode.

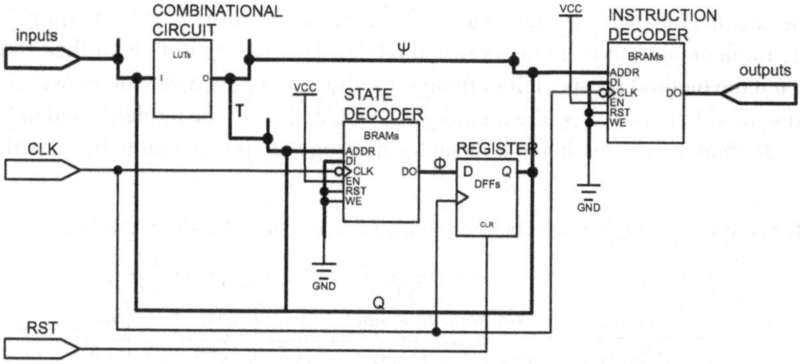

Fig. 8.18 The schematic diagram of PAY_0 Mealy FSM

This is only one possible architecture that could be obtained after architectural decomposition with the application of multiple encoding. The architecture depends on which parameter(s) is(are) encoded and which parameter is used as partition set [14]. The presented architecture yields very good results [13], but the gain in hardware resources consumption is strongly dependent on the characteristics of the control algorithm [3]. It means that architecture and method of encoding should be chosen individually for each algorithm.

8.5 Summary

The increasing complexity of the digital systems and novel advanced technologies have become essential for application of efficient logic synthetic methods as well as digital design and mapping tools. However, the commercial tools usually take into account some trade-off between computational complexity and the quality of the physical implementation of the projects [45, 48]. The diversity of implementations requires the use of advanced methods of logic synthesis, however most existing systems on digital market does not implement such solutions using the outdated synthesis methods. The advancement of logic synthesis methods, the complexity of procedures, and novel digital structures leads to the situation that only computer-aided systems developed mainly at academic research centers are able to support digital design [6, 9, 13, 14, 16, 27, 29, 31, 38, 39, 46, 47].

The idea presented in the chapter is targeted at complex concurrent behavior specified with statechart, which is finally implemented in modern programmable devices equipped with memory blocks and configurable logic. The transformation from statechart diagram into FSM model is carried with symbolic formal methods efficiently implemented by means of Binary Decision Diagrams. Both synthesis strategies (ROM-based scheme with address modifier and architectural decomposition) consume hardware resources to different extent.

ROM-based method uses only memory blocks. Table 8.4 presents the gain in memory bits obtained with functional decomposition scheme and implementation

in the architecture with address modifier and memory (Fig. 8.12b) in comparison
to the implementation without address modifier. Decreasing ratio is of the order of
tens and the method is especially efficient for more complex behavior. However, the
address modifier, as well as the memory content, can easily be implemented in logic
cells [38] making the method universal for heterogeneous programmable structures.

Table 8.4 FSM synthesis results before and after introduction of address modifier

name	#in	#out	#q	#cube	before #bit	after #bit	gain %	dec. ratio
Garage	6	3	4	49	7168	1664	77	4.3
TVRemoteControl	8	5	4	55	36864	6912	81	5.3
SimpleReactor	10	15	8	986	6029312	294912	95	20.4
ReactorWithWagon (Fig. 8.4)	10	9	6	263	983040	26112	97	37.6

$$gain = \frac{\#bit_{before} - \#bit_{after}}{\#bit_{before}} \cdot 100\% \qquad decreasing\ ratio = \frac{\#bit_{before}}{\#bit_{after}}$$

Architectural decomposition uses both, memory and configurable logic. According to Figure 8.16, blocks CC and Y are mapped into memory, and blocks P and
RG are implemented in LUTs. Table 8.5 presents the gain in LUTs obtained with
the application of architectural decomposition in comparison to standard FSM synthesis method which is well known VHDL template utilizing only LUTs [43]. The
results were obtained using Xilinx tool with default settings.

Table 8.5 FSM synthesis results with standard method and as PAY_0 Mealy FSM

name	Standard method #LUT	#FF	PAY_0 #LUT	#FF	#BRAM	LUT gain %	dec. ratio
Garage	82	14	10	4	2	88	8.2
TVRemoteControl	102	12	28	4	2	73	3.6
SimpleReactor	1965	163	472	13	23	77	4.2
ReactorWithWagon (Fig. 8.4)	423	33	46	6	5	89	9.2

$$gain = \frac{\#LUT_{std\_method} - \#LUT_{PAY_0}}{\#LUT_{std\_method}} \cdot 100\% \qquad decreasing\ ratio = \frac{\#LUT_{std\_method}}{\#LUT_{PAY_0}}$$

The idea of functional decomposition which is the base for address modifier concept can be applied to any function [31]. Its application to functions implemented
in blocks P, Y and CC can bring further reductions in hardware resources, not only
memory bits, but also in configurable logic.

Decomposition methods proved their efficiency for the latest programmable devices. It seems that combining architectural synthesis with functional ROM-based decomposition is very promising for logic controller design, especially for structures equipped with embedded memory blocks.

Acknowledgements. The research was partly financed from budget resources intended for science in 2010 – 2013 as an own research project No. N N516 513939.

References

1. Adamski, M., Barkalov, A.: Architectural and sequential synthesis of digital devices. University of Zielona Góra, Zielona Góra (2006)
2. Baranov, S.I.: Logic synthesis for control automat. Kluwer Academic Publishers, Boston (1994)
3. Barkalov, A., Titarenko, L.: Logic synthesis for FSM-based control units, Lecture Notes in Electrical Engineering, vol. LNEE, vol. 53. Springer, Heidelberg (2009)
4. Bazydło, G., Adamski, M.: Logic controllers design from UML state machine diagrams. Czasopismo Techniczne: Informatyka z. 24, 3–18 (2008) (in Polish)
5. Borowik, G., Rawski, M., Łabiak, G., Bukowiec, A., Selvaraj, H.: Efficient logic controller design. In: Fifth International Conference on Broadband and Biomedical Communications (IB2Com), pp. 1–6 (December 2010), doi:10.1109/IB2COM.2010.5723633
6. Borowik, G.: Improved state encoding for FSM implementation in FPGA structures with embedded memory blocks. Electronics and Telecommunications Quarterly 54(1), 9–28 (2008)
7. Borowik, G., Kraśniewski, A.: Trading-off error detection efficiency with implementation cost for sequential circuits implemented with FPGAs. In: Moreno-Díaz, R., Pichler, F., Quesada-Arencibia, A. (eds.) EUROCAST 2011, Part II. LNCS, vol. 6928, pp. 327–334. Springer, Heidelberg (2012)
8. Borowik, G., Łuba, T., Tomaszewicz, P.: On memory capacity to implement logic functions. In: Moreno-Díaz, R., Pichler, F., Quesada-Arencibia, A. (eds.) EUROCAST 2011, Part II. LNCS, vol. 6928, pp. 343–350. Springer, Heidelberg (2012)
9. Brzozowski, J.A., Łuba, T.: Decomposition of boolean functions specified by cubes. Journal of Multi-Valued Logic and Soft Computing 9, 377–417 (2003)
10. Buchenrieder, K., Pyttel, A., Veith, C.: Mapping statechart models onto an FPGA-based ASIP architecture. In: EURO-DAC 1996, pp. 184–189 (September 1996), doi:10.1109/EURDAC.1996.558203
11. Bukowiec, A.: Synthesis of FSMs based on architectural decomposition with joined multiple encoding. International Journal of Electronics and Telecommunications 58(1), 35–41 (2012), doi:10.2478/v10177-012-0005-7
12. Bukowiec, A., Barkalov, A., Titarenko, L.: Encoding of internal states in synthesis and implementation process of automata into FPGAs. In: Proceedings of the Xth International Conference on the Experience of Designing and Application of CAD Systems in Microelectronics, CADSM 2009, pp. 199–201. Ministry of Education and Science of Ukraine and Lviv Polytechnic National University, Lviv, Publishing House Vezha & Co., Polyana, Ukraine (2009) ISBN: 978-966-2191-05-9
13. Bukowiec, A.: Synthesis of finite state machines for FPGA devices based on architectural decomposition. Lecture Notes in Control and Computer Science, vol. 13 (2009)

14. Bukowiec, A., Barkalov, A.: Structural decomposition of Finite State Machines. Electronics and Telecommunications Quarterly 55(2), 243–267 (2009)
15. Czerwiński, R., Kania, D.: State assignment for PAL-based CPLDs. In: Wolinski, C. (ed.) Proc. of 8th Euromicro Conference on Digital Systems Design, Architectures, Methods and Tools, Porto, Portugal, August 30-September 3, pp. 127–134. IEEE Computer Society (2005), doi:10.1109/DSD.2005.71
16. Doligalski, M.: Behavioral specification diversification for logic controllers implemented in FPGA devices. In: Proceedings of the 9th Annual FPGA Conference - FPGAworld 2012, Stockholm, Sweden, pp. 6:1–6:5. ACM (2012), doi:10.1145/2451636.2451642
17. Drusinsky, D., Harel, D.: Using statecharts for hardware description and synthesis. IEEE Transaction on Coputer-Aided Design 8(7), 798–807 (1989), doi:10.1109/43.31537
18. Drusinsky-Yoresh, D.: A state assignment procedure for single-block implementation of state chart. IEEE Transaction on Coputer-Aided Design 10(12), 1569–1576 (1991), doi:10.1109/43.103506
19. Gajski, D.D., Vahid, F., Narayan, S., Gong, J.: Specification and design of embedded systems. Prentice-Hall, Inc., Upper Saddle River (1994)
20. Gomes, L., Costa, A.: From use cases to system implementation: statechart based co-design. In: Proceedings of 1st ACM & IEEE Conference on Formal Methods and Programming Models for Codesign, MEMOCODE 2003, Mont Saint-Michel, France, pp. 24–33. IEEE Computer Society Press (2003), doi:10.1109/MEMCOD.2003.1210083
21. Harel, D.: Statecharts: A visual formalism for complex systems. Science of Computer Programming 8(3), 231–274 (1987), doi:10.1016/0167-6423(87)90035-9
22. Hartmanis, J., Stearns, R.E.: Algebraic structure theory of sequential machines. Prentice-Hall, New York (1966)
23. Henson, M.A.: Biochemical reactor modeling and control. IEEE Control Systems Magazine 26(4), 54–62 (2006), doi:10.1109/MCS.2006.1657876
24. I-Logix Inc., 3 Riverside Drive, Andover, MA 01810 U.S.A.: STATEMATE Magnum Code Generation Guide (2001)
25. Jenkins, J.H.: Designing with FPGAs and CPLDs. Prentice Hall, Upper Saddle River (1994)
26. Karatkevich, A.: Deadlock analysis in statecharts. In: Forum on Specification on Design Languages (2003)
27. Łabiak, G.: From UML statecharts to FPGA - the HiCoS approach. In: Proceedings of Forum on Specification & Design Languages, FDL 2003, pp. 354–363. Frankfurt am Main (September 2003)
28. Łabiak, G.: The use of hierarchical model of concurrent automaton in digital controller design. University of Zielona Góra Press, Poland, Zielona Góra (April 2005) (in Polish)
29. Łabiak, G.: From statecharts to FSM-description – transformation by means of symbolic methods. In: 3rd IFAC Workshop Discrete-Event System Design, DESDes 2006, Poland, pp. 161–166 (2006), doi:10.3182/20060926-3-PL-4904.00027
30. Łabiak, G., Borowik, G.: Statechart-based controllers synthesis in FPGA structures with embedded array blocks. Intl Journal of Electronic and Telecommunications 56(1), 11–22 (2010), doi:10.2478/v10177-010-0002-7
31. Łuba, T., Borowik, G., Kraśniewski, A.: Synthesis of Finite State Machines for implementation with programmable structures. Electronics and Telecommunications Quarterly 55(2) (2009)
32. de Micheli, G.: Symbolic design of combinational and sequential logic circuits implemented by low-level logic macros. IEEE Transactions on CAD CAD-5(4), 597–616 (1986), doi:10.1109/TCAD.1986.1270230

33. de Micheli, G.: Synthesis and optimization of digital circuits. McGraw-Hill Higher Education (1994)
34. de Micheli, G., Brayton, R.K., Sangiovanni-Vincentelli, A.: Optimal state assignment for finite state machines. IEEE Transactions on CAD CAD-4(3), 269–284 (1985), doi:10.1109/TCAD.1985.1270123
35. Minato, S.: Binary decision diagrams and applications for VLSI CAD. Kluwer Academic Publishers, Boston (1996)
36. Ramesh, S.: Efficient translation of statecharts to hardware circuits. In: Twelfth International Conference on VLSI Design, pp. 384–389 (January 1999), doi:10.1109/ICVD.1999.745186
37. Rawski, M., Selvaraj, H., Łuba, T.: An application of functional decomposition in ROM-based FSM implementation in FPGA devices. Journal of Systems Architecture 51, 424–434 (2005), doi:10.1016/j.sysarc.2004.07.004
38. Rawski, M., Tomaszewicz, P., Borowik, G., Łuba, T.: 5 Logic synthesis method of digital circuits designed for implementation with embedded memory blocks of FPGAs. In: Adamski, M., Barkalov, A., Węgrzyn, M. (eds.) Design of Digital Systems and Devices. LNEE, vol. 79, pp. 121–144. Springer, Heidelberg (2011)
39. Sasao, T.: On the number of LUTs to realize sparse logic functions. In: Proc. of the 18th International Workshop on Logic and Synthesis, Berkeley, CA, U.S.A., July 31-August 2, pp. 64–71 (2009)
40. Szecówka, P.M., Pedzińska-Rżany, J., Wolczowski, A.R.: Hardware approach to artificial hand control based on selected DFT points of myopotential signals. In: Moreno-Díaz, R., Pichler, F., Quesada-Arencibia, A. (eds.) EUROCAST 2009. LNCS, vol. 5717, pp. 571–578. Springer, Heidelberg (2009)
41. Villa, T., Sangiovanni-Vincentelli, A.: NOVA: state assignment of finite state machines for optimal two-level logic implementation. IEEE Transactions on CAD 9(9), 905–924 (1990), doi:10.1109/43.59068
42. Yang, S.: Logic synthesis and optimization benchmarks User Guide Version 3.0. Tech. rep., Microelectronics Center of North Carolina, P.O. Box 12889, Research Triangle Park, NC 27709 (1991)
43. Zwolinski, M.: Digital system design with VHDL. Prentice Hall (2004)
44. Zydek, D., Selvaraj, H., Borowik, G., Luba, T.: Energy characteristic of processor allocator and network-on-chip. Journal of Applied Mathematics and Computer Science 21(2), 385–399 (2011), doi:10.2478/v10006-011-0029-7
45. Altera: Embedded Memory in Altera FPGAs (2010),
    http://www.altera.com/technology/
    memory/embedded/mem-embedded.html
46. FSMdec: FSMdec Homepage (2012),
    http://gborowik.zpt.tele.pw.edu.pl/node/58
47. HiCoS: HiCoS Homepage (2012),
    http://www.uz.zgora.pl/%7eglabiak
48. Xilinx: Block RAM (BRAM) Block (v1.00a), San Jose (August 2004),
    http://www.xilinx.com/support/documentation/
    ip_documentation/bram_block.pdf

Chapter 9
Virtualization from University Point of View

Tomasz Babczyński, Agata Brzozowska, Jerzy Greblicki, and Wojciech Penar

Abstract. In this chapter we present some problems connected to cloud computing and in particular to virtualization. At university cloud computing is on the one hand interesting tool for researchers and on the other very interesting topic to research. We present network simulator with nodes done in virtual environment (in this case virtualization is a tool) and scheduling algorithm for cloud computing (in this case cloud computing is object of our researches).

9.1 Introduction

In this chapter we present some problems connected to cloud computing and in particular to virtualization. At university cloud computing is on the one hand interesting tool for researchers and on the other very interesting topic to research. We present network simulator with nodes done in virtual environment (in this case virtualization is a tool) and scheduling algorithm for cloud computing (in this case cloud computing is object of our researches).

First part of this chapter we present simple problem of task scheduling as a one of the main parts of Cloud Computing (CC). We understand this process as mapping users tasks to appropriate resources to execute. Since we use virtual machines (VM) to hide physical resources form user the critical problem is how to schedule this machines (sub tasks) in given resources (ie. host machines). Tasks scheduling algorithms for CC systems are discussed in literature [11, 8, 5, 4]. The most authors use mathematical model without consideration of many problems with desktop virtualization. We consider a new model for tasks scheduling of VM deployment for business and educational purposes. We also introduce genetic algorithm [6] for task scheduling in CC.

Tomasz Babczyński · Agata Brzozowska · Jerzy Greblicki · Wojciech Penar
Institute of Computer Engineering, Control and Robotics,
Wrocław University of Technology, 11-17 Janiszewskiego Street, 50-372 Wrocław, Poland
e-mail: {tomasz.babczynski,agata.brzozowska,jerzy.greblicki,
        wojciech.penar}@pwr.wroc.pl

© Springer International Publishing Switzerland 2015   153
R. Klempous and J. Nikodem (eds.), *Innovative Technologies in Management and Science*,
Topics in Intelligent Engineering and Informatics 10, DOI: 10.1007/978-3-319-12652-4_9

In the second part of this chapter we would like to present some aspects of simulation of network as a most important part of distributed systems [10]. Distributed systems, private clouds, in many cases reflect the natural character of the network infrastructure of the units, such as ATM networks, or multiple branches in different cities. Such solutions are much harder to design, configure and management than centralized systems, but of course, increase the reliability of the system. In case of failure of one node, the others continue working. It affects not only proper working, but also in a significant way to delay the whole system. We should therefore pay special attention to the design and testing. The idea of simulation studies of behavior of a selected object by its model. Strong interest in such a way to study and solve problems are the result of the primary advantages of simulation, namely:

- minimization of the costs (the possibility of making rapid and radical modification, without having to increase financial resources)
- saving time needed for installation / modification of infrastructure
- saving space and energy
- the possibility of testing scenarios, which are hard to emulate using real hardware such as damage or attacks on the network.

There have been many programs (simulators), simulating the operation of computer networks. Such programs are often created "by researchers for researchers," so it is not surprising that the main emphasis is on performance analysis capabilities and not on simplicity.

9.2 Cloud Computing

Cloud computing is a new processing model. In the last few years, the term Cloud is increasingly popular in both the commercial use and as a scientific term. "While 2010 may have been the year of cloud talk, 2011 is the year of cloud action." said Dr Ajei Gopal, Executive Vice President Products and Technology Group, CA Technologies. "Cloud computing is changing the way business operates and the way the IT functions. IT is no longer a back-room activity, but rather a major business enabler shaping every aspect of how an organization operates". Because of a large number of possible applications, it is hard to create one sufficiently wide definition.

Cloud computing allows to share resources, software, and information over the Internet. But the most important is the provision of services by external organizations. Cloud Computing is therefore primarily to provide services. The technology that allows it relates to the concept of virtualization. We understand Cloud as resources pool and above that virtualization.

Another definition describes Cloud as three things: thin client, grid computing and utility computing. Grid links computers to large infrastructure, and may or may not be in the Cloud. Utility computing gives us possibility of paying only for actual consumption (such as gas, electricity and services).

Very important ideas are also "access on demand" and "pay as you go". On demand provisioning means that client defines parameters of interest and pays for exactly what is ordered. If his needs grow 'on demand', he may increase the resources by buying more storage or performance and the additional service will be added on the fly and without stopping working applications and virtual machines.

9.2.1   Types of Cloud

There are two main divisions of the Clouds. The first is the distinction between Cloud public, private and hybrid. Public Cloud shares resources via the Internet and charges for utilization. Public Cloud is very easy to scale. This hybrid gives centralization and standardization, increased security but also flexibility, thanks to possible access via VPN clients. Private Cloud use internal network and gives security and surveillance. The second divides the models of Clouds on Saas, PaaS, IaaS and Haas. Software as a Service (SaaS) rely on software distribution. The application is stored and made available to users through the Internet. This eliminates the need for installing and running applications on the client side. SaaS shifts the responsibilities of management, updates, support from the consumer to the supplier. Platform as a Service (PaaS) is a service, which gives virtual environment for work. This service is primarily aimed at developers. Infrastructure as a Service (IaaS) delivers computer infrastructure as a service.

9.2.2   Benefits and Concerns

"The rise of the cloud is more than just another platform shift that gets geeks excited. It will undoubtedly transform the information technology (IT) industry, but it will also profoundly change the way people work and companies operate. It will allow digital technology to penetrate every nook and cranny of the economy and of society, creating some tricky political problems along the way." [12]

Benefits of managing in the Cloud are: reduced cost, increased storage and flexibility. Other advantages are the independence of the architecture and portability. Reduced cost because limitation of need to purchase a license (or licenses cheaper and better used), no need of installation and software administration. Possibility of scaling the system without changing the infrastructure. Do not purchase the equipment in case of an emergency temporary demand. Pay as you go - we do not pay for equipment when it is not fully exploited. It is also important that small business does not need to employ technical support

We have to ask how Cloud Computing will affect software development? Is migration of existing mechanisms possible? How to program large-scale distributed systems? How to work with large disk arrays? From the economical point of view we have even more doubts. How to estimate total cost of ownership Cloud? Which business model choose? How to provide availability and continuity of service. And the biggest conserve touches data security issues. Are my files safe? Is there any law regulation for Cloud? And finally: Are we ready for Cloud?

9.2.3   Trends and Examples

The top research subjects are: using old hardware with Cloud, migration of existing systems to Cloud, standardization of management, connecting mobile devices to Cloud, fear of huge data leakage and desktop virtualization. There are also many open questions concerning energy consumption and heat emitted by data centres, investment costs and even sociological aspects of using Cloud Computing.

The most famous existing Cloud System is made by Google. Huge number of users of such services as Google Documents, Google Mail or Google Buzz, everyday confirm bright future of Cloud Computing.

9.3   Scheduling Algorithm for Virtualization Tasks

In this section We present the optimization problem in the Cloud Computing architecture. This is a real problem that grow up at our University.

We show its transformation to the particular packing problem and meta-heuristic methods engaged to find an appropriate problem solution.

Task scheduling is one of the main parts of Cloud Computing (CC). We understand this process as mapping users tasks to appropriate resources to execute. Performance of whole cloud computing depends on this process. Since we use virtual machines (VM) to hide physical resources form user the critical problem is how to schedule this machines (sub tasks) in given resources (ie. host machines). The most authors of tasks scheduling algorithms for CC systems discussed in literature [11, 8, 5, 4] use mathematical model without consideration of many problems with desktop virtualization. Our new model for tasks scheduling of VM deployment for business and educational purposes is presented here. Especially in heterogeneous systems this task is an NP-complete. In recent years CC systems have been rapidly developed. Undeniably for this type of system it is necessary to design new algorithms for management including scheduling. One of the problem that needs analysis is scheduling in for desktop virtualization CC systems. Task scheduling is one of the classic problem of optimization. Solution is not obvious, because that problem is NP complete. Especially in a heterogeneous system, such as CC it is not trivial.

Most of the proposed solutions use meta heuristic algorithms. We also introduce genetic algorithm [1] for task scheduling in CC. Genetic Algorithms use techniques inspired by natural evolution (i.e. crossover, mutation) and are widely used for optimization problems.

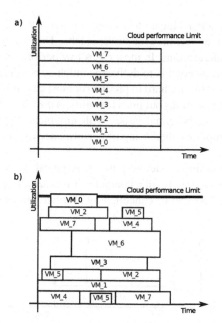

Fig. 9.1 Typical workloads for business (a) and research (b) applications

9.3.1   Problem Description

There is a need of novel design and analysis methodology for private Cloud for education. Methods from business/industry do not apply to educational area (i.e. problem of performance parallelism). Figure 9.1 shows two different life cycles of virtual machines. The first one can be found in companies like banks, where all employees work for eighth hours per day. The second is typical for research data centres, where clients tasks are not known a priori. None of those models fits academical environment.

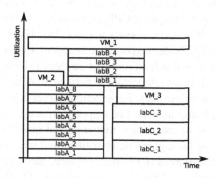

Fig. 9.2 Example workload for education

Example of workload in education is shown on figure 9.2. System has two goals. One is to deliver virtual machines for students labs. This machines are easy to predict, have typical configurations. In most cases low performance is required but inaccessibility is critical. Second is to deliver as much as possible other machines for research and teachers. This are hard to predict (in most case immediate deployment) and have various configurations but inaccessibility is not critical. Cloud System will face more problems like potential issues with untypical hardware (remote USB, GPU computing), temporary increasing of performance (parallel work in groups of students).

We need new scheduling model because of untypical tasks, which are not compute intensive and do not have regular character. Tasks are also performance flexible, but not time flexible. Task can be cancelled by user but not delayed.

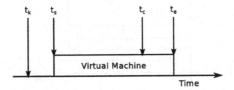

Fig. 9.3 Live cycle of virtual machine

In university applications of planning tasks for the CC, there is a need to provide multiple virtual machines (VM) with different performance requirements, for the given time intervals. Academic Cloud is consisted of always available computers (host machines). Number of host machines and their specification are known. For some VM required working hours are known and are strictly given by university staff. Each of VM is also described by memory size and CPU performance. Figure 9.3 presents typical live cycle of an virtual machine, where: Tk-request time, Ts-start time, Te-stop time, Tc-cancel time (unexpected event). Users connect to Cloud to access virtual machines from terminals. In our model to schedule deployment of virtual machines we need information such as: the total number of machines and for each of them: minimum required CPU performance, minimum required memory size, working time. It is also necessary to place this tasks on a system time-line. Starting time can be precisely defined or it can be fluent. Furthermore, it is assumed that the hardware resources of at least one host machine is larger than those required by the virtual machine. It is also important to noticed, that in academic applications we have several other dependencies. For example server VM for classes has to be started before other VM start and can't be stopped when machines are still running. In light of this, we have to create new mathematical model for academic Cloud. In considered system we have given number of virtual machines. Each of them is described by required resources (CPU performance, memory). It is assumed that these requirements may be different for each machine. On the other side there is a CC system consisting of multiprocessor computers (host machines) with defined parameters. It is assumed that each virtual machine must be deployed on a single host. It is not allow for a virtual machine to use resources from several hosts.

To solve this problem it is necessary to create its mathematical model. Model from [2], unfortunately, can not be applied to the problem of tasks scheduling of desktop virtualization. Authors use the worst-case calculation time. In the considered problem, this approach does not apply because the deployment time (starting and end time) is pre-set. Therefore, the key problem in the task of planning the distribution of virtual machines is to ensure the specified performance at a specified time (i.e. working hours, time classes for students). In light of this we can not handle the worst-case calculation time. In our model VM must be deployed on a single host. It is not allow for a virtual machine to use resources from several hosts. Host changes are prohibited when machine is running. It is possible to run two and more than two virtual machines on one host. Host resources used by machine are reserved for it for whole machine's run time and can be used only by this machine. When virtual machine is stopped, resources reserved for it are immediately released and can be assigned to the next task. Schedule algorithm deploys each virtual machine on selected host where it will be executed. Starting time is also specified in this process.

It is easy to notice that the problem described above is of the big complexity due to the great number of restrictions of different kinds, for the first. One may observe that our problem looks like a bin-packing problem whereas it is also similar from other points of view to the vehicle routing problem. Hosts may be treating like a truck (vehicles) and virtual machines are the objects to be carried. A special source of difficulty comes from the time relations between particular virtual machines. These make our problem similar to project management task and can make it unfeasible i.e. unable to be realized from time to time. These properties carry on the evident observation that it is impossible to write down our problem using only mathematical formulas. We propose new mathematical model for this problem:

n - the number of virtual machines
m - the number of blades in the host
T - time limit: $t \in [0,T]$
p_i - advisable performance of VM i, $i=1,2,...,n$
q_i - minimal necessary performance of VM i, $i=1,2,...,n$
c_i - the priority of VM i, $i=1,2,...,n$
t_i - the starting time of VM i, $i=1,2,...,n$
v_i - the ending time of VM i, $i=1,2,...,n$
f_i - feasible performance of the j-th blade, $j=1,2,...,m$

Goal: find the feasible mapping (x_i, y_i), where:

$x_i \in 0,1,2,...,m$, $i=1,2,...,n$
$q_i \leq y_i \leq p_i$, $i=1,2,...,n$.

Goal function - case A

$$F(x,y) = \sum_{i=0}^{n} c_i l(x_i) \frac{y_i}{p_i}, F(x,y) \to max \qquad (9.1)$$

n - the number of virtual machines
m - the number of the hosts
pi - advisable performance of i-th VM, i=1,2,...,n
ci - the priority of i-th VM, i=1,2,...,n

Decision variables: (xi, yi):

xi $\in$ 0,1,2,?m, i=1,2,...,n : CASE A
qi $\leq$ yi $\leq$ pi, i=1,2,...,n.

Goal function - case B

$$F(x,y) = \sum_{i=0}^{n} c_i l(x_i) \frac{y_i}{p_i}, F(x,y) \to max \qquad (9.2)$$

n - the number of virtual machines
m - the number of the hosts
pi - advisable performance of i-th VM, i=1,2,...,n
ci - the priority of i-th VM, i=1,2,...,n

Decision variables: (xi, yi):

xi $\in$ 1,2,...,m, i=1,2,...,n : CASE B
qi $\leq$ yi $\leq$ pi, i=1,2,...,n.

So, to obtain a useful tool for the problem solving we put into the motion meta-heuristic approach [3]. Namely, for our purposes we reworked an optimization procedures based on the idea of Genetic Algorithm and Harmony Search Algorithm. Both approaches need for the very beginning an appropriate coding procedures to conduct necessary operators easy. To check the feasibility of reworked solution we used same special projections from the space of genotypes to the space of phenotypes based on the Baldwin Effect. Between others, this idea made our algorithms more flexible for the future needs i.e. restrictions.

Overview of our scheduling analysis system for educational CC is presented on figure 9.4. We collect statistical data from our real laboratory network including processor load, number of processors, memory and storage usage. Than we prepare example workloads for CC system and we analyse scheduling algorithms and cloud configuration. It is also possible to do reconfiguration of analysed system and recalculate nodes utilisation.

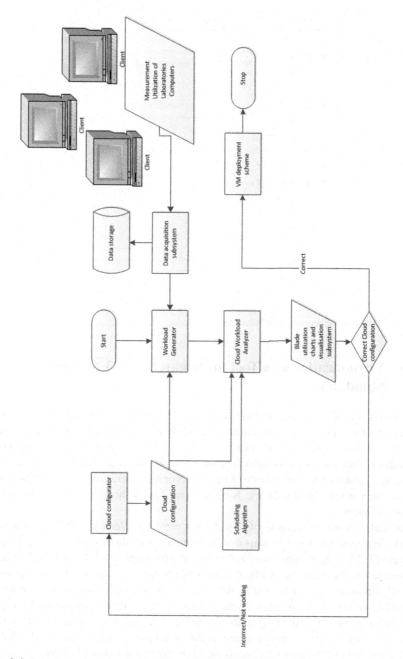

Fig. 9.4 Scheduling system for Cloud Computing in education

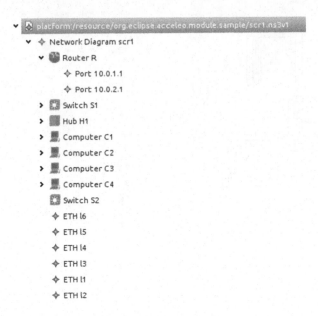

Fig. 9.5 Network schema in XMI formatk

9.4 Virtualization as a Tool in Research – NS3 Network Simulator

Virtualization of computers and network environment is also interesting as a tool in research. Some part of our researches requires heterogenous enviroment of several nodes connected with network. Our previous researches with help of NS-3 network simulator were extended to virtualization of nodes in cloud. Such a modification gave as a great tool for verification of behaviors or real operation systems (Windows, Linux etc.) in simulated network. Next sections covers short example of simulated network system.

In earlier work the authors of this chapter were presented possibilities of a network simulator NS-3 and how its use for simulation of distributed data acquisition system [1, 2]. System consists of a client station (manager) and several data sources connected to the network. A client station sends queries to data sources that send a reply. Inquiries are connected to subjects. Every fixed quantity of data source is specified levels of competence in each subject. These levels determine the probability of a correct answer. The above publication appeared ample opportunities simulator NS-3. The results obtained in the simulation can be considered reliable and comparable with the experiments made on the real network [3, 7, 9].

One purpose of this work is to create a user-friendly graphical interface for the selected network simulator. The program should allow the visualization of network topology and determine its parameters and simulation options. The most important here is clarity and simplicity in use.

In this chapter we present tools based on Eclipse [17] and NS-3. Eclipse is a description of the transformation of the network model to XMI format as fit for use by the network simulator NS-3. So it is a description of the second part of the process performance analysis to real-time distributed system using the methodology of "model" (called *model-driven*).

Eclipse is a very powerful environment that supports many plug-ins, that extend its functionality. The translation process uses a model possibility of Acceleo package [12]. Model is converted into a form understandable for the network simulator NS-3. The choice of a simulator from a wide variety of market is justified below. Metamodel used in the work was prepared using the package Eclipse Modeling Framework (EMF), and network models are created using a graphical editor created using the package GEF (Graphical Editing Framework) and GMF (Graphical Modeling Framework). Editor makes possible a graphic presentation of a distributed system, including network diagram and the applications running on the nodes. To transform the network model Acceleo package is used. It compliant with OMG group for the transformation of models for text. These transformations are used as the last stage of transformation in the MDD methodology for artefacts like program code, report, specification or other text or hypertext document. For our work, artefact produced during the transformation is the program code for the simulator NS-3. The method of its creation and fragments of conversion scripts will be shown in next paragraph.

Choosing the right network simulating tool is essential for the success of the project. The following criteria of his choice: - Free Software for commercial purposes. - Open source project, providing the possibility of any modification. - The project supports the standard output file format. The simulator should also enable: - testing packet networks, - simulations in real time, - virtualization or some other method of starting the actual application, - analysis of the traffic at each node of the simulated system. NS-3 simulator meets all these criteria. Allows a real-time simulation, and it is necessary to start the actual application on simulated network nodes. In addition, the availability of source code makes modifications possible. There is significant support for virtualizations methods.

Network Simulator, also popularly known as the Berkeley network simulator is discrete event simulator, which means that the time increases by leaps and bounds, but his increments are variable (sequence of events is here more important than the actual passage of time). Work on the project started in 1989 at the University of Berkeley. Since 2006 a third version of the simulator has developed, which is not backwards compatible with enjoying the confidence of users of Ns-2. One of most interesting features of this simulator is that user nodes can be implemented as a virtualized operating systems including Windows XP, Windows 7 and Linux based.

The project is supported by many research institutions including both academic and commercial. The whole is open source (GNU GPLv2) created by researchers for researchers. Aided design and testing of network packet protocols, supports the simulation of TCP, UDP, CBR, FTP, HTTP, routing, multicast, and wireless and wired networks and large-scale networks. Produces output files standard format, eg. pcap files (the ability to analyze with tcpdump or Wireshark) and mobility scripts compatible with the NS-2. This allows the study in real time. NS-3 code is written in C++and documented by the doxygen documentation generator (API documentation). There are also electronic documentation in the form of manual and tutorial[13, 14, 15, 17, 16, 18]. Support is available on the newsgroup, and wiki pages. The main objectives of the project is research and education networks. Unfortunately, user documentation for NS-3 is still incomplete and it induces both experienced and new users to use in their research Ns-2. There have been efforts to create a simple to use tools that make it easier to work with the simulator NS-3. Such a tool should first give graphical user interface, which is missing in the simulator NS-3. In 2009, at the University of Strasbourg NS-3-generator[16] was created. The application allows entering the network topology in graphical form, and writing the model in XML and finally code generation - the NS-3 in C++or Python. It was created using Qt, so it has a modern layout, is also very intuitive to use. However, the project is not developed for almost a year. It also has many drawbacks and limitations such as: applications installed on the nodes are not visible, can not be modified or even removed, the inability to modify the simulations settings, network addresses are assigned automatically. The work interferes with the lack of highlight the currently selected element and the lack of undo option. Also co-author of this chapter during the implementation of the thesis developed using WindowsForms technology Generator v.2 program, which has similar features as described above and, unfortunately, similar defects.

Example network shown in Figure 9.6 consists of nodes and links. At the point where nodes contact with links there are ports. Nodes can perform a functionality of devices such as PCs, routers, hubs, or switches. It is possible to implement PC nodes as a virtualized systems. Each of these devices may have any number of ports. Each port is described by specification calls. And connections can be of three types: Ethernet, ADSL, and custom (user defined). However, presented diagram is only a graphical presentation of the network model, which is written in a language XML Metadata Interchange (XMI).

Network shown in Figure 9.4 is an Ethernet network. Therefore, the ports are characterized by IP and MAC addresses. Default IP address is 10.0.0.0 and the default mask is equals 24. The network consists of seven nodes, one of them acts as a router, a switch, a hub, while the others are computers. One two computers (C3 and C4) are installed applications (UdpEchoServer and UdpEchoClient). During the simulation, these applications communicate with each other. The server runs on port nine, waiting for a connection from the client. Computer nodes were virtualized in VirtualBox connected together with internal network.

Example network diagram were created wit Acceleo software. In that project we can create a mtl file. Code written in Acceleo Model to Text Transformation Language (MTL). The following code creates a text file with the same name as the project and the cc extension.

```
[comment encoding = UTF-8 /]
[module generate('http:///ns3v1')/]
[template public generate(aNetworkDiagram :
  NetworkDiagram)]
```

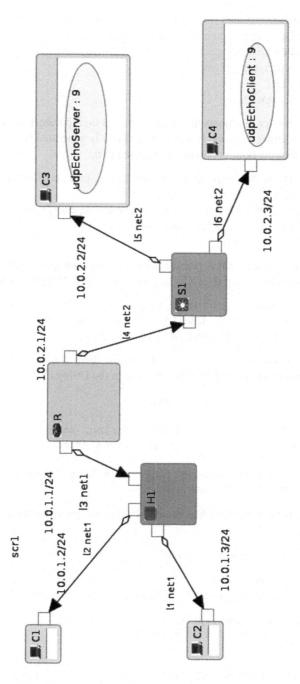

Fig. 9.6 Graphic representation of network

```
[comment @main /]
[file (aNetworkDiagram.name.concat('.cc'), false,
 'UTF-8')]
   //Kod programu
[/file]
[/template]
```

Acceleo can refer to objects in the diagram and their parameters. The following code checks whether the node is a switch, and if it is generated Ns-3 code, responsible for creating the nodes container connected to it.

```
[if (aNetworkDiagram.node->filter(Switch)->size()>0)]
  NodeContainer nodesSwitch;
  nodesSwitch.Create ([aNetworkDiagram.node->
      filter(Switch)->size()/]);
[/if]
```

To automate translation it is possible to write your own queries and write limitations in the model by using the OCL language (Object Constraint Language).

```
[query public returns_port(arg : Port) : Port =
    Sequence {arg.downLink.from, arg.upLink.to}->
    reject(oclIsInvalid())->first() /]
```

Another possibility is to use Java code. Below there is a cod calling resolv _ip (String add, int mask) metod, which is located in the Utility class.

```
[itt.IP.resolv_ip(itt.NetMask)/];
```

```
[query public resolv_ip(addr : String, mask : Integer):
    String = invoke ('org.eclipse.acceleo.module.sample.
    files.Utility',
    'resolv_ip(java.lang.String, java.lang.Integer)',
    Sequence {addr , mask})/]
```

Java code is in a separate file named Utility.java. Its contents can be seen on listing below.

```
package org.eclipse.acceleo.module.sample.files;

public class Utility {
  public String resolv_ip(String addr, Integer mask){
  if(mask == 24)   {
      String[] a=addr.split("\\.");
      return a[0]+"_"+a[1]+"_"+a[2];
      }
  else{
      String[] a=addr.split("\\.");
      return a[0]+"_"+a[2];
      }
   }
}
```

This section presents a method of translating the network diagram XMI format to source code of simulator NS-3 in the Eclipse [13, 14, 15, 17, 16, 18]. As part of further work is planned to expand the language of other types of networks, especially wireless (due to the high possibility of NS-3 in this field). And to extend the opportunity to write user applications, including user-friendly way (by combining ready-made modules) .

The present work concerns the simulation of distributed systems, private clouds. The tool is to streamline the tedious process of preparing the simulation, which is a first in-depth analysis of the structure of the test network and the tasks performed by it, and then at such a selection of the simulator and its parameters so that the obtained results best reflect the actual behavior of the tested system. This approach to solving the problem is very cost effective even for economic reasons.

Modern simulators give us the impression of working on a real network, such as through a real IP addresses, packets containing the bytes of real networks, work in real time. Thanks to this often time-consuming construction of complex models of system under consideration becomes unnecessary. Moreover, it becomes a real chance of testing their own applications, including those compiled.

NS-3 is a tool designed mainly for research and education. However, the process of learning to use it is relatively long, in the absence of basic programming skills impossible. The need for the creation of graphical user interface is indicated by multiple users. The answer to these needs is to be produced under this work program. It facilitates the introduction of the network structure to the computer. The program generates C code, compatible with the simulator NS-3.

9.5   Conclusions

In this chapter we have presented virtualization for university point of view. We consider virtualization as a interesting tool for our research purposes and also topic of our researches.

The presented work concerns the simulation of distributed systems, private clouds. The tool is to streamline the tedious process of preparing the simulation, which is a first in-depth analysis of the structure of the test network and the tasks performed by it, and then at such a selection of the simulator and its parameters so that the obtained results best reflect the actual behavior of the tested system. This approach to solving the problem is very cost effective even for economic reasons.

Modern simulators give us the impression of working on a real network, such as through a real IP addresses, packets containing the bytes of real networks, work in real time. Thanks to this often time-consuming construction of complex models of system under consideration becomes unnecessary. Moreover, it becomes a real chance of testing their own applications, including those compiled. Virtualization gave user a great opportunity to test real operating systems and applications in simulated environment.

Ns-3 is a tool designed mainly for research and education. However, the process of learning to use it is relatively long, in the absence of basic programming skills impossible. The need for the creation of graphical user interface is indicated by multiple users. The answer to these needs is to be produced under this work program. It facilitates the introduction of the network structure to the computer. The program generates C code, compatible with the simulator Ns-3.

Virtual machines deployment for educational purposes has other than business type restrictions. We have checked several deployment schemes and developed pre-scheduling reduces performance algorithm specially for students labs. Our future plans concern improvement of pre-scheduling and approximation of mathematical model coefficient from real system.

References

1. Babczyński, T., Brzozowska, A.: Performance evaluation of real distributed systems in simulated network. In: Borzemski, L. (ed.) Information Systems Architecture and Technology. New Developments in Web-Age Information Systems, pp. 125–135. Oficyna Wydawnicza Politechniki Wrocławskiej, Wrocław (2010)
2. Babczyński, T., Penar, W.: Badania wydajnościowe rozproszonego systemu akwizycji danych. In: Mazur, Z., Huzar, Z. (eds.) Modele i Zastosowania Systemów Czasu Rzeczywistego, pp. 71–80. Wydawnictwa Komunikacji i Łącznosci, Warszawa (2008)
3. Banks, J.: Discrete-Event System Simulation. Prentice Hall (2000)
4. Cagan, J., Shimada, K., Yin, S.: A survey of computational approaches to three-dimensional layout problems. Computer Aided Design 34(8), 597–611 (2002)
5. Garg, S.K., et al.: Environment-conscious scheduling of HPC applications on distributed Cloud-oriented data centers. J. Parallel Distrib. Comput. (2010)
6. Greblicki, J., Kotowski, J.: Analysis of the properties of the Harmony Search Algorithm carried out on the one dimensional binary knapsack problem. In: Moreno-Díaz, R., Pichler, F., Quesada-Arencibia, A. (eds.) EUROCAST 2009. LNCS, vol. 5717, pp. 697–704. Springer, Heidelberg (2009)
7. Hartmann, A.K.: A Practical Guide To Computer Simulation. World Scientific (2009)
8. Kotowski, J.: The use of the method of illusion to optimizing the simple cutting stock problem. In: Proc. MMAR 2001, 7th IEEE Conference on Methods and Models in Automation and Robotics, vol. 1, pp. 149–154 (2001)
9. Rahman, M.A., Wang, F.Z.: Network modelling and simulation tools, Londyn (2007)
10. Tanenbaum, A.S., Van Steen, M.: Distributed Systems: Principles and Paradigms. Prentice-Hall, Inc. (2002)
11. Zhao, C., et al.: Independent Tasks Scheduling Based on Genetic Algorithm in Cloud Computing. In: Proc. 5th Int. Conf. on Wireless Comm., Net. and Mobile Computing, WiCom 2009 (2009)
12. A survey of corporate IT. Let it rise, The Economist (2011)
13. Eclipse Modeling Framework Project (EMF), http://www.eclipse.org/modeling/emf (access May 01, 2011)
14. Graphical Editing Framework (GEF), http://www.eclipse.org/gef (access May 01, 2011)

15. Graphical Modeling Project (GMP),
    `http://www.eclipse.org/modeling/gmp` (access May 01, 2011)
16. WebPage of Acceleo, `http://www.eclipse.org/acceleo/` (access April 20, 2011)
17. WebPage of Eclipse, `http://www.eclipse.org` (access April 20, 2011)
18. WebPage of NS-3, `http://www.nsnam.org/` (access April 20, 2011)
19. Tutorial Ns-3, `http://www.sfc.wide.ad.jp/tazaki/distfiles/ns3/ns-3-tutorial-en.pdf` (access April 21, 2011)

Chapter 10
Metaheuristic Algorithms for the Quadratic Assignment Problem: Performance and Comparison

Andreas Beham, Michael Affenzeller, and Erik Pitzer

Abstract. The quadratic assignment problem is among the harder combinatorial optimization problems in that even small instances might be difficult to solve and for different algorithms different instances pose challenges to solve. Research on the quadratic assignment problem has thus focused on developing methods that defy the problem's variety and that can solve a vast number of instances effectively. The topic of this work is to compare the performance of well-known "standard" metaheuristics with specialized and adapted metaheuristics and analyze their behavior. Empirical validation of the results is performed on a highly diverse set of instances that are collected and published in form of the quadratic assignment problem library. The data in these instances come from real-world applications on the one hand and from randomly generated sources on the other hand.

10.1 Introduction

10.1.1 Quadratic Assignment Problem

The quadratic assignment problem (QAP) was introduced in [15] and is a well-known problem in the field of operations research. It is the topic of many studies, treating the improvement of optimization methods as well as reporting successful application to practical problems in keyboard design, facility layout planning and re-planning as well as in circuit design [10, 13, 7]. The problem is NP hard in general and, thus, the best solution cannot easily be computed in polynomial time. Many

Andreas Beham · Michael Affenzeller · Erik Pitzer
Heuristic and Evolutionary Algorithms Laboratory
School of Informatics, Communication and Media
University of Applied Sciences Upper Austria, Research Center Hagenberg
Softwarepark 11, 4232 Hagenberg, Austria
e-mail: {andreas.beham,michael.affenzeller,
        erik.pitzer}@fh-hagenberg.at

© Springer International Publishing Switzerland 2015   171
R. Klempous and J. Nikodem (eds.), *Innovative Technologies in Management and Science*,
Topics in Intelligent Engineering and Informatics 10, DOI: 10.1007/978-3-319-12652-4_10

different optimization methods have been applied, among them popular metaheuristics such as tabu search [20, 14] and genetic algorithms [8].

The problem can be described as finding the best assignment for a set of facilities to a set of locations so that each facility is assigned to exactly one location which in turn houses only this facility. An assignment is considered better than another when the flows between the assigned facilities have to be hauled over smaller distances.

The QAP is also a generalization of the traveling salesman problem (TSP). Conversion of a TSP can be achieved by using a special weight matrix [15, 17] where the flow between the "facilities" is modeled as a ring that involves all of them exactly once. The flow in this case can be interpreted as the salesman that travels from one city to another.

More formally the problem can be described by an $N \times N$ matrix W with elements w_{ik} denoting the weights between facilities i and k and an $N \times N$ matrix D with elements d_{xy} denoting the distances between locations x and y. The goal is to find a permutation π with $\pi(i)$ denoting the element at position i so that the following objective is achieved:

$$\min \sum_{i=1}^{N} \sum_{k=1}^{N} w_{ik} \cdot d_{\pi(i)\pi(k)} \tag{10.1}$$

The complexity of evaluating the quality of an assignment according to Eq. (10.1) is $O(N^2)$, however several optimization algorithms move from one solution to another through small changes, such as by swapping two elements in the permutation. These moves allow to reduce the evaluation complexity to $O(N)$ and even $O(1)$ if the previous qualities are memorized [20]. Despite changing the solution in small steps iteratively, these algorithms can, nevertheless, explore the solution space and interesting parts thereof quickly. The complete enumeration of such a "swap" neighborhood contains $N * (N-1)/2$ moves and, therefore, grows quickly with the problem size. This poses a challenge for solving larger instances of the QAP.

The QAP can also be used to model cases when there are more locations than facilities and also when there are more facilities than locations. In these cases dummy facilities with zero flows or dummy locations with a high distance can be defined.

QAPLIB

The quadratic assignment problem library [4] (QAPLIB) is a collection of benchmark instances from different contributors. According to their website[1], it originated at the Graz University of Technology and is now maintained by the University of Pennsylvania, School of Engineering and Applied Science. It includes the instance descriptions in a common format, as well as optimal and best-known solutions or lower bounds and consists of a total of 137 instances from 15 contributing sources which cover real-world as well as random instances. The sizes range from 10 to 256 although smaller instances are more frequent. All 103 instances between 12 and 50

---

[1] http://www.seas.upenn.edu/qaplib/

have been selected for this study with the exception of esc16f, which does not specify any flows.

10.1.2   Literature Review

Research on the QAP has a longer history, but is still very active. A few selected publications shall be briefly described here to give the reader an overview on the research efforts that have been conducted.

The first successful metaheuristic for the QAP was robust taboo search (RTS) and has been introduced in [20]. This algorithm addresses one of the main weaknesses of the standard tabu search (TS) algorithm. In tabu search it can often happen that the search trajectory returns to the same solution over and over. This cycling behavior is due to its rather deterministic nature and the attempt of Taillard was to randomize the tabu tenure and therefore make the trajectory more robust. Taillard also noted that some instances of the QAP, namely the els19, proved to be difficult to be solved. The search would get stuck in a certain sub-region of the search space and would be unable to escape as it would never climb high enough to leave it. He described a simple diversification strategy that would choose to make diversifying moves at certain times which were enough to solve the problem.

The genetic local search (GLS) algorithm has been introduced in [16]. It is a hybrid metaheuristic that combines elements of a genetic algorithm with local search. The authors note that local search is already able to solve the QAP quite well if it is applied multiple times, as will also be shown here. They developed a new crossover operator which they named *DistancePreservingCrossover*. It is a highly disruptive crossover that only preserves the common alleles and randomizes all other genes. The common alleles are then fixed and the local search is applied to find a new local optima for the other genes.

The performance of iterated local search was analyzed in [19]. In that work Stützle tested several strategies of an iterated local search algorithm. The local search is a first-improvement local search that would make use of so called "don't look bits" which limit the neighborhood of the local search to the interesting parts. He also uses concepts of variable neighborhood search in that he varies the degree of perturbation of the solution from which the local search is restarted after it has landed in a local optima. Finally, he hybridizes this algorithm with an evolution strategy that achieves very competitive results.

A survey and a little bit of history on solutions to some quadratic assignment problem instances is given in [9]. Drezner also proposed several new problem instances, which have however not found their way into the QAPLIB yet.

The topic of fitness landscape analysis on the quadratic assignment problem has initially been attempted in [17]. Merz and Freisleben described several measures such as autocorrelation and correlation length to measure landscape ruggedness and fitness distance correlation (FDC) to visualize the correlation between solution similarity (to the optimum) and quality. Problems are easier to solve when it holds that the more similar a solution becomes to the optimum the better its quality. This can lead the search to the optimum instead of deceiving it.

The problem ruggedness has also been analyzed in [3]. Angel introduced a ruggedness coefficient and stated that the QAP appears to be a rather smooth problem which is beneficial for local search. Autocorrelation values recently have also been calculated exactly in [5]. Chicano et al obtain the autocorrelation function analytically instead of empirically.

In the following chapter, results and a comparison between different algorithms will be given on several instances of the quadratic assignment problem taken from the QAPLIB. The quality differences will be given on a scale between the best-known respectively optimal quality and an instance specific average. The reason is that the typically used relative difference between the obtained solution's fitness $f(s)$ and the best known solution f^*, given as $(f(s) - f^*)/f^*$, cannot be used to compare algorithm performance over multiple problem instances as a given instance might appear to be solved well, when actually only the range between an average solution's quality and the optimum is narrow. Therefore, we use the average fitness of the problem instance $\hat{f}$ to normalize this ratio and obtain a *scaled difference* as $(f(s) - f^*)/(\hat{f} - f^*)$. It is shown in [5] that $\hat{f}$ can be computed exactly for any instance of the QAP. The comparison of effort between the algorithms has been normalized to the computational task to evaluate a solution. The swap move that is often used can be calculated in O(N) and even O(1) [20]. If it is calculated in O(1) there are just 4 operations necessary, otherwise there are $4 * N$ operations required. The evaluation of a full solution requires N^2 operations. In this chapter *solution evaluation equivalents* denotes the amount of full solution evaluations taking into account that move evaluations need less operations. If not otherwise mentioned all results in this chapter are computed as an average of 20 repetitions of the same configuration.

10.2 Trajectory-Based Algorithms

Trajectory-based algorithms attempt to move from one solution to the next through a number of smaller or larger steps. Different strategies exist to guide the movement. The most often used neighborhood in the QAP is the so called swap2 neighborhood where two indices in the permutation are randomly selected and exchanged. For a QAP solution this means that two facilities are placed in the location of the respective other facility. This move creates only a small change. All of the trajectory-based algorithms in this chapter make use of this type of move.

10.2.1   Local Search

Local search basically comes in two variants:

- **best-improvement:** Considers a number of moves at the same time. It attempts to make only the move with the maximum performance gain. It is sometimes also referred to a steepest or gradient-descent heuristic. If no more improving moves can be made the search knows to have reached a local optimum.
- **first-improvement:** Considers moves one after another and attempts to make each move that would improve the current solution. It does not follow the steepest gradient, but in some cases where the neighborhood is much larger than the number of steps to the local optimum first-improvement might be quicker to converge. It knows to have reached a local optimum when it has tried all moves and none constituted an improvement.

The performance of local search depends to a large degree on the number and quality of local optima that would attract the search. Typically a quadratic assignment problem contains many local optima that are scattered over the entire search space. Only some instances lead the search into similar local optima. In Figure 10.1 the similarity of local optima is shown for two selected instances. In the `els19` instance a randomly started local search often converges into similar local optima, while in the `bur26a` instance there are much more and scattered local optima.

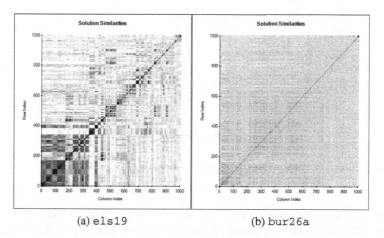

(a) `els19`   (b) `bur26a`

Fig. 10.1 Similarities of 1,000 randomly sampled local optima with each other for two selected problem instances. The darker an entry in this matrix the more similar two local optima are.

In Figure 10.2 it can be observed that the similarity of local optima reduces with the problem size which can be expected given the increase of the search space. However, the similarities change at a slightly slower rate (n^c) than search space ($n!$).

Maximum similarity reduces to about 20% for problems of size 50 which means that for a given local optimum there is at least one other optimum out of 999 that is 20% similar. Average similarity compares the similarity among all other local optima found.

As the results in Figure 10.3 show, small instances can be very deceptive for a local search. Although the problems belonging to the rou family are still small enough that the trajectory leads to the best solution the average trajectory ends up about 20% away from the optimum and 45% in the worst observed case. In general, the average quality of local optima varies greatly between the instances as well as the effort to reach them. It is also interesting to note that in the lipa-b instances the average fitness of local optima is rather high. The effort to find such an optimum using a best-improvement local search based on the swap2 neighborhood is steadily increasing with the problem size. While it takes on average 286 solution evaluation equivalents to find a local optimum in the rou family it takes about 4,837 solution evaluation equivalents for the wil50 problem instance. Although such a trend exists, the difficulty of instances of the quadratic assignment problem cannot only be judged purely by problem size from the point of view of a local search algorithm.

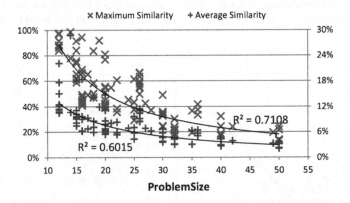

Fig. 10.2 Correlation of local optima similarity and problem size

The local search is often repeated to find a different and better local optimum in the next descent. Such a repetition might be performed from a completely random solution, or from a slight perturbation of the current local optimum in which case it is called an iterated local search (ILS). Table 10.1 shows the results from simple iterated first- and best-improving local search strategies. The algorithms were run with a maximum of 10,000 repeated starts. The search was stopped as soon as it reached the known optimum or best-known quality and the effort is given in the average number of evaluated solution equivalents. The search is restarted from a solution where on average about $\frac{1}{4}$ of the current solution is randomly perturbed. Tests were also conducted with a simple repeated local search with comparable, but slightly

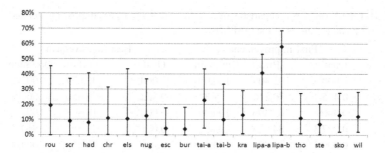

Fig. 10.3 Distribution of local optima with respect to their fitness values. The graphic shows the minimum, average, and maximum fitness for local optima found from 10,000 initial random samples.

worse results. The main difference between a basic iterated local search and a repeated local search is the required effort. Starting over from completely new random solutions requires considerable more effort to descend into a local optimum. When starting closer to the just found local optimum finding a new local and possibly also the global optimum is faster. In the case of iterated best-improvement local search the effort is about 180% higher when repeating from a new random solution, in first-improvement it is still about 52% higher. Nevertheless, there is the possibility that the trajectory gets stuck in a certain region. In our case however the strategy was very simple as the search always continued from the newly found local optimum and the scramble operation is able to modify even the whole permutation, although with only a very small probability.

It is also interesting to compare lipa-a and lipa-b instances with tai-a and tai-b. While the -b instances generally appear to be easier to solve lipa-b also requires a lot less effort, the effort for tai-b however remains roughly the same. Another interesting observation is that els of which the only instance is els19 requires less effort in the best-improvement variant. This is noteworthy as it is the most significant out of three groups that showed this behavior. Overall, the roughly similar quality, but in general reduced effort of the first-improvement search makes it much more attractive to be used.

10.2.2   Simulated Annealing

In contrast to local search simulated annealing allows the trajectory to worsen with a certain probability. This counters the drawback of local search algorithms that get stuck in local optima as they are missing a strategy for escaping it. Simulated annealing uses a temperature parameter that governs the probability of accepting a worsening move. An often used cooling schedule is the exponential cooling where the probability to move from solution c to solution x in the case of a fitness

Table 10.1 Performance comparison of a simple first- and best-improving iterated local search on several instances of the QAPLIB

Instance	First-improvement			Best-improvement		
	Quality	Effort	Optimum	Quality	Effort	Optimum
bur	0.000%	105,409	100.0%	0.000%	232,074	100.0%
chr	0.475%	570,816	59.3%	0.322%	1,091,356	66.4%
els	0.000%	62,496	100.0%	0.000%	14,808	100.0%
esc	0.037%	84,323	95.9%	0.025%	241,453	96.8%
had	0.000%	5,324	100.0%	0.000%	22,223	100.0%
kra	0.066%	790,856	80.0%	0.127%	3,091,380	70.0%
lipa-a	10.083%	1,592,687	61.3%	8.695%	4,678,133	68.8%
lipa-b	0.000%	10,777	100.0%	0.000%	41,671	100.0%
nug	0.011%	205,228	96.0%	0.014%	535,476	95.7%
rou	0.023%	297,756	83.3%	0.074%	558,576	81.7%
scr	0.000%	28,307	100.0%	0.000%	145,315	100.0%
sko	0.803%	3,558,891	7.5%	0.845%	12,069,038	7.5%
ste	0.212%	2,126,174	33.3%	0.259%	7,214,283	28.3%
tai-a	3.843%	1,700,668	37.8%	4.195%	4,270,270	36.1%
tai-b	0.032%	1,506,763	65.6%	0.043%	4,827,457	60.6%
tho	0.407%	2,103,004	37.5%	0.522%	6,299,726	30.0%
wil	0.693%	4,473,760	0.0%	0.661%	15,415,083	0.0%
0.849%	**713,103**	**76.4%**	**0.812%**	**2,063,833**	**76.6%**

deterioration is $p(x,c) < e^{\frac{f(c)-f(x)}{T}}$. As the temperature parameter T is reduced over time the search is less and less likely to accept large uphill moves.

As can be seen in the results of Table 10.2 the performance depends to a large degree on the start temperature. If the initial temperature is not chosen high enough the search will converge too quickly, if however the initial temperature is raised the search may converge into better regions of the search space. However, raising the temperature too far is not beneficial as the search would then only perform a random walk and waste CPU time. In the table the initial temperatures have been chosen according to the range between the average quality and the best-known quality. The search should accept with a probability of 1% a solution that worsens the current solution by 1%, 4%, 16%, or 32% of the fitness range. Note that in this case the best-known quality has been used, but if that is not available one could also use a suitable lower bound.

There are further parameters in simulated annealing that govern the speed of the cooling as well as the initial and final temperature. Cooling speed of 1 represents a linear cooling schedule while 0.01 and 0.00001 represent more rapid cooling schedules that would quickly prevent the acceptance of uphill moves. As shown in Figure 10.4 experiments regarding the variation of these parameters have shown that only the starting temperature has a significant influence on the final solution quality.

Detailed results are given in Table 10.3 for the configuration that worked best overall. It is interesting to see that simulated annealing is having difficulties in

Table 10.2 Performance of simulated annealing averaged over various configurations with respect to the chosen start temperature

T_0	Quality	Effort	Optimum
1%	7.155%	3,666,504	24%
4%	1.594%	2,507,105	53%
16%	**0.503%**	**1,760,454**	**71%**
32%	0.596%	2,158,207	69%

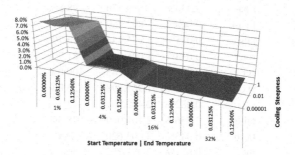

Fig. 10.4 Parameter range test for simulated annealing shows that start temperature has the largest effect on the performance. In this case the results are averaged over all repetitions of all tested instances.

solving the els19 instance to optimality. On the other hand the algorithm has no trouble solving many of the other instances. The seemingly difficult lipa-a instances are solved with much more ease finding the optimum in all, but one case. Instance families that simulated annealing is not as well suited as a simple iterated local search are bur, had, nug, scr, and tai-b. In these instances the performance is worse despite sometimes a much larger effort that is required in obtaining good solutions. On the other hand simulated annealing is able to solve chr, kra, rou, sko, ste, tai-a, tho, and wil better than the simple iterated local search.

10.2.3   Tabu Search

Like simulated annealing tabu search also has a strategy for escaping local optima, which forces the search to make a move every time which is the best move that can currently be made. Naturally, this would likely return the search to the previous solution in the succeeding iteration so tabu search includes a memory that prevents to revert a move that has been made recently. This memory thus allows the search to explore greater parts of the search space.

The so called "standard" tabu search includes a simple memory that remembers the previously assigned location. Every move is declared *tabu* if it would reassign

Table 10.3 Performance of simulated annealing applied to several instances of the QAPLIB

		16%	
Instance	Average	Effort	Optimum
bur	0.247%	3,528,124	41.9%
chr	0.213%	2,137,507	67.1%
els	1.531%	4,901,637	15.0%
esc	0.000%	220,586	100.0%
had	0.183%	1,483,900	82.0%
kra	0.106%	1,544,613	85.0%
lipa-a	0.336%	839,576	98.8%
lipa-b	0.000%	802,769	100.0%
nug	0.035%	955,344	94.7%
rou	0.034%	1,095,646	81.7%
scr	0.000%	637,245	100.0%
sko	0.291%	1,962,898	25.0%
ste	0.082%	2,402,201	43.3%
tai-a	2.756%	2,101,099	38.3%
tai-b	0.345%	3,003,620	53.1%
tho	0.081%	2,072,757	50.0%
wil	0.406%	1,968,492	5.0%
0.379%		**1,619,005**	**73.2%**

a location to a facility that it had already been assigned to within the last n iterations. Nevertheless sometimes a move should be made even though it is tabu, such as when it would find a new best solution. The so called aspiration condition ensures that such moves are considered nevertheless. In the "standard" tabu search the aspiration condition typically reconsiders moves that would find a new best solution. However, more advanced aspiration conditions exist that take the quality of a certain move into account and would undo a move if it improved the quality from the last time it was done. This can happen when other parts of the solution have changed significantly. Tabu search typically has one important parameter that governs the time that moves are kept tabu: the tabu tenure. The longer a certain move is kept tabu the less likely the search returns and the more it will diversify. However, if it is too long the search might be prevented to intensify and cannot explore the region around local optima. With less possible moves in the neighborhood to choose from the behavior approaches that of random search.

The results shown in Table 10.4 indicate the performance of standard tabu search applied to the already mentioned problem instances. The algorithm generally shows good search behavior, but it has problems to solve some instances such as from the bur, els, had, and tai-b family. As Taillard states in [20] in certain instances certain moves are never considered and thus the search is trapped in a sub-region of the search space. One possibility to counter this would be to increase the tabu tenure and thus force the search to diversify, however as already mentioned if the tenure becomes too large the search is only diversifying and will not descend and explore

Table 10.4 Performance of a standard tabu search

| | MaxIter = 2,000 | | | | MaxIter = 100,000 | |
| | tenure = N | | tenure = 2*N | | tenure = 2*N | |
Instance	Quality	Optimum	Quality	Optimum	Quality	Optimum
bur	2.772%	6.9%	2.341%	10.0%	2.121%	20.6%
chr	2.984%	16.8%	1.568%	23.2%	0.254%	69.6%
els	8.972%	0.0%	8.678%	10.0%	6.407%	10.0%
esc	0.320%	92.1%	0.214%	92.4%	0.015%	99.7%
had	1.539%	48.0%	1.265%	53.0%	1.069%	57.0%
kra	2.934%	11.7%	2.267%	21.7%	0.205%	90.0%
lipa-a	16.055%	40.0%	12.571%	55.0%	0.000%	100.0%
lipa-b	1.429%	97.5%	1.426%	97.5%	0.000%	100.0%
nug	0.821%	53.0%	0.644%	56.7%	0.057%	97.0%
rou	1.168%	36.7%	0.443%	68.3%	0.000%	100.0%
scr	1.570%	45.0%	0.100%	80.0%	0.000%	100.0%
sko	1.115%	2.5%	1.693%	0.0%	0.458%	12.5%
ste	1.326%	10.0%	1.152%	5.0%	0.169%	43.3%
tai-a	4.785%	10.6%	4.375%	19.4%	1.704%	46.7%
tai-b	8.258%	11.9%	7.559%	10.0%	7.325%	20.6%
wil	1.925%	0.0%	2.054%	0.0%	0.567%	0.0%
	2.991%	**39.2%**	**2.410%**	**44.1%**	**1.080%**	**69.6%**

interesting local optima. Therefore Taillard proposed the robust taboo search (RTS) that would randomize the tabu tenure to counter the probability the search stagnates. He also added a diversification strategy to explore still unseen parts of the search space. This algorithm has thus two main parameters the tabu tenure that governs the random variable from which the actual tenures are drawn and the aspiration tenure that defines a fixed number of iterations after diversifying moves are being made. These two parameters guide the intensifying and diversifying behavior of the search trajectory. Smaller tabu tenures lead to more intensification of close local optima, while larger tabu tenures will aim to diversify more. In a similar way a small aspiration tenure leads to more frequent diversification while larger aspiration tenures allow to intensify the search longer.

An extensive parameter study was conducted where for each of the 102 instances 90 configurations were tested. The configurations were chosen by combining each of 9 different tabu tenures (25, 50, 100, 150, 200, 300, 400, 600, and 800) with 10 different aspiration tenures (100, 500, 750, 1000, 1500, 2000, 3500, 5000, 7500, and 10000). The algorithm was allowed a total of 100,000 iterations. The averaged results over all instaces from the study can be seen in Figure 10.5 while the results of the average best configuration (tabu tenure of 200 and aspiration tenure of 7,500) are given in Table 10.5.

The robust taboo search is generally a very well working metaheuristic for problems of this size, however the parameter configuration is a bit harder. Both the tabu

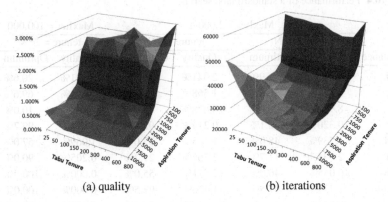

(a) quality   (b) iterations

Fig. 10.5 Parameter range test for robust taboo search shows that best results are achieved with a moderate tabu tenure and a higher aspiration tenure. Each point in the grid was evaluated 20 times by performing the algorithm on all 102 instances.

tenure and the aspiration tenure have a high and sometimes interdependent influence on the performance. Good parameter settings need to be chosen, otherwise the search will perform worse. Unfortunately it is rather difficult to give a good figure for tabu search's effort. The ability to evaluate certain moves in $O(1)$ as described in [20] greatly increases the influence of other elements on the search effort. The effort is calculated by computing each move evaluation as 4 operations in Table 10.5.

10.2.4 Variable Neighborhood Search

Variable neighborhood search (VNS) is a further trajectory-based algorithm with another strategy to escape local optima. Instead of trying to simply "move on" as would tabu search do or escape by probability as would simulated annealing do the variable neighborhood search attempts to perform a set of changes to the current solution in the hope of perturbing it to a degree that allows different local optima to be reached. It is similar to an iterated local search, but it uses a more advanced strategy to continue the search. The strength of the modifications increases with the number of attempts: at first only small changes are made, but as the search remains unsuccessful in finding a better local optimum the strength of the change is increased. This creates a well working combination of fast local search behavior with a mutation-based diversification strategy. The performance of VNS is also shown in Table 10.5 in which the algorithm was run for 500 full cycles of shaking operators. The results shown are the averages of those runs. It finds comparable solutions to a first-improvement local search although it has been given much less time.

Table 10.5 Performance of the robust taboo search (RTS) and the variable neighborhood search (VNS)

Instance	RTS				VNS		
	Quality	Iterations	Effort	Optimum	Quality	Effort	Optimum
bur	0.112%	37,964	73,008	88.8%	0.000%	29,753	100.0%
chr	0.173%	36,709	69,834	75.4%	0.379%	210,424	59.3%
els	0.000%	23,805	45,103	100.0%	0.000%	14,187	100.0%
esc	0.025%	6,092	11,802	97.1%	0.021%	25,067	97.9%
had	0.228%	17,730	33,435	90.0%	0.000%	4,213	100.0%
kra	0.176%	37,061	71,690	91.7%	0.708%	548,332	53.3%
lipa-a	0.000%	11,712	22,833	100.0%	13.795%	630,936	50.0%
lipa-b	0.000%	1,106	2,156	100.0%	0.000%	32,010	100.0%
nug	0.000%	5,362	10,274	100.0%	0.103%	123,707	84.7%
rou	0.001%	9,244	17,543	98.3%	0.247%	144,587	70.0%
scr	0.000%	1,752	3,314	100.0%	0.000%	21,900	100.0%
sko	0.109%	58,454	114,423	55.0%	0.667%	1,259,354	20.0%
ste	0.015%	41,464	80,625	91.7%	0.097%	461,140	70.0%
tai-a	1.042%	50,626	98,112	62.8%	4.025%	597,374	27.8%
tai-b	0.116%	41,990	81,548	73.8%	0.048%	251,736	86.9%
tho	0.030%	51,952	101,211	57.5%	0.596%	843,418	17.5%
wil	0.137%	86,947	170,417	40.0%	0.702%	1,684,106	0.0%
	0.159%	**24,836**	**47,895**	**86.6%**	**1.033%**	**244,711**	**75.1%**

It applies swap2, swap3, scramble, inversion, insertion, translocation, and translocation-inversion as shaking operators in this order [2].

10.3   Population-Based Algorithms

Population-based algorithms attempt to make use of a larger number of solutions that are evolved over time. Often in optimization it is beneficial to explore the search space more rigorously in order to find better solutions. There are a number of strategies which involve various degrees of replacement and elitism, that control how the algorithms would discard a solution and instead accept a new one. For instance, in a genetic algorithm with generational replacement the newly evolved population always fully replaces the old population, even if it was of worse quality. Often only one single best individual is retained from the old population. This strategy requires the crossover operator to combine relevant genetic information of high quality individuals. In the offspring selection genetic algorithm however, the replacement strategy puts more pressure towards fulfilling this requirement: The search accepts new children only if they outperform their respective parents. This strategy often works better than a standard genetic algorithm, however in each generation many more children are produced and the convergence is slowed. At the same time this allows a more thorough exploration of the search space which should yield higher quality results.

In contrast to trajectory-based algorithms, the introduction of a population should benefit in those cases when the fitness landscape has many different basins of attraction [18]. This does correlate with the number of local optima, but not fully as local optima might also be quite close to another and form valleys of attraction (it is believed that e.g. the traveling salesman problem does generate such a "big-valley" landscape in which good solutions are also quite similar to each other [6]). Population-based search, at least in the beginning, is able to discover multiple such basins and explore them simultaneously. Trajectory-based algorithms always only explore to the root of one basin at a time and then have to find their way to the next basin.

As was observed, the behavior of the standard tabu search on the els19 resulted in a search trajectory that got stuck in a confined part of the search space. The same problem instance is however comparably easy to solve for a genetic algorithm. However, as mentioned, population-based algorithms are only successful if the distributed information on relevant genes can be combined in one chromosome through the process of survival of the fittest and crossover. Especially when solving the QAP it can be observed that this process is not always successful.

10.3.1 Genetic Algorithm

The performance of the genetic algorithm depends to a large part on whether the crossover operator can combine the relevant genes from multiple individuals in a new offspring. However, as can be seen in a parameter study involving various crossover operators which are generally deemed suitable, various mutation operators, and mutation probabilities the genetic algorithm population cannot effectively converge in optimal regions of the search space. The search converges prematurely and stagnates at a certain level to the known optimum with much of the diversity lost. At this point crossover is not relevant anymore and only mutation may introduce new alleles into the population. Table 10.6 shows the average performance of standard genetic algorithm with a population size of 500, and full generational replacement. For the tests with roulette-wheel selection the algorithm was run only for 5,000 generations, but achieved on average better results than tournament selection run for 10,000 generations. This shows that the lower selection pressure is able to maintain genetic diversity longer which increases the chance of combining it in a single solution.

As Figure 10.6 shows the genetic diversity in the population can be lost as early as generation 30 if selection pressure is very high. After the search has converged the continued evolution is mainly driven by random mutations with a low probability and therefore highly inefficient. Using a selection operator that excerts less selection pressure such as roulette-wheel in this case shows that convergence can be

Table 10.6 Performance of a standard genetic algorithm with partially matched crossover (PMX) [11], swap2 manipulation and 15% mutation probability

Instance	5-Tournament			Roulette		
	Quality	Effort	Optimum	Quality	Effort	Optimum
bur	1.221%	4,897,518	1.9%	1.338%	2,494,196	0.6%
chr	5.569%	4,765,084	4.6%	4.245%	2,212,262	15.4%
els	0.460%	1,508,229	70.0%	0.743%	2,400,715	5.0%
esc	1.307%	1,006,638	80.3%	1.089%	543,318	80.0%
had	1.505%	2,645,654	48.0%	0.984%	1,126,099	60.0%
kra	9.021%	4,990,500	0.0%	7.745%	2,493,571	1.7%
lipa-a	38.429%	4,928,337	1.3%	3.695%	2,388,408	5.0%
lipa-b	50.978%	4,435,188	11.3%	6.410%	2,158,538	17.5%
nug	6.122%	4,791,288	4.0%	4.488%	2,314,742	9.0%
rou	10.605%	4,990,500	0.0%	6.970%	2,137,933	15.0%
scr	4.701%	4,243,156	15.0%	3.702%	2,197,747	21.7%
sko	10.211%	4,990,500	0.0%	7.410%	2,495,500	0.0%
ste	5.342%	4,990,500	0.0%	7.326%	2,495,500	0.0%
tai-a	17.875%	4,962,855	0.6%	4.003%	2,408,973	3.9%
tai-b	3.334%	4,305,529	13.8%	2.773%	2,203,563	12.5%
tho	8.795%	4,990,500	0.0%	6.684%	2,495,500	0.0%
wil	9.227%	4,990,500	0.0%	0.459%	2,495,500	0.0%
	8.737%	**3,978,419**	**20.4%**	**7.315%**	**1,969,243**	**23.1%**

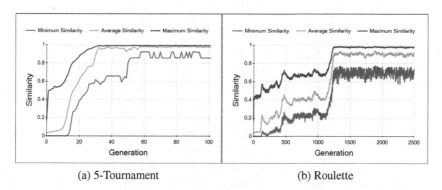

(a) 5-Tournament   (b) Roulette

Fig. 10.6 The chart displays the average similarities in the population. It can be seen that the genetic diversity in the population of the GA is lost by generation 30 already when optimizing the lipa30a instance with high selection pressure. If the selection pressure is less the population takes longer to converge allowing the search to explore more of the search space.

prolonged for many generations. Selection pressure generally refers to the ratio of selecting better solutions to selecting worse solutions.

10.3.2 Offspring Selection Genetic Algorithm

The offspring selection genetic algorithm (OSGA) [1, 2] is an advanced and more robust variant of a genetic algorithm. Instead of relying on the undirected stochastic nature of the crossover and mutation operators it introduces another selection criteria for the newly created offspring. If the performance of the offspring surpasses that of its parents it is admissible to the next population, if however it is worse than the parents (the better parent) the offspring is discarded. This places a variable evolutionary pressure on the population forcing it to produce better and better offspring. The visible effect is that the OSGA performs much better than the standard GA in both convergence speed and quality. Still, the performance is not entirely satisfying and it is likely possible that even in this case the crossover operator is not able to combine the relevant genetic information. The performance of OSGA is given in Table 10.7. The SuccessRatio was set to 0.5 and the mutation probability was set to 25%. The algorithm selects one parent with tournament selection and a group size of 3 and the other parent randomly. In this case a variant was used where the individuals that remained after the SuccessRatio was filled were not selected from the pool of unsuccessful individuals, but were randomly selected from the previous population.

Table 10.7 Performance of the offspring selection genetic algorithm (OSGA)

Instance	Quality	Effort	Optimum
bur	0.479%	388,609	11.3%
chr	1.932%	416,956	27.1%
els	0.000%	63,000	100.0%
esc	1.033%	117,199	84.1%
had	0.590%	167,030	62.0%
kra	4.742%	462,953	0.0%
lipa-a	31.173%	626,594	12.5%
lipa-b	32.062%	557,218	47.5%
nug	1.987%	378,197	23.3%
rou	3.114%	406,432	20.0%
scr	0.742%	217,933	63.3%
sko	6.876%	669,170	0.0%
ste	2.332%	500,843	0.0%
tai-a	11.810%	528,069	14.4%
tai-b	1.319%	376,965	27.5%
tho	4.523%	477,328	0.0%
wil	4.787%	608,935	0.0%
****	**5.013%**	**369,409**	**34.3%**

As expected from previous applications of OSGA [2] the algorithm could greatly improve the results compared to the standard genetic algorithm. However, a few exceptions are interesting to note regarding instances of type lipa-a and lipa-b, as well as `tai-a` and `wil`. In these instances OSGA was able to find the optimal or best known solution more often, however the average quality is worse than in the standard GA. It can also be seen that these are instances where the standard genetic algorithm with tournament selection also struggled much more. A selection scheme with less selection pressure seems to be the more favorable approach, meaning that the search needs to diversify with a much greater degree. A similar observation can be made when the results from standard tabu search and robust taboo search are compared.

10.4 Hybrid Algorithms

The goal of hybridizing algorithms is to combine good properties of a number of algorithms in one strategy. From the previous experience on performance indicators it can be derived that trajectory-based algorithms with their step-wise local search behavior appear to work very well given a certain diversification strategy.

The intent of this chapter is to show and analyze possible performance improvements by combining population-based and trajectory-based algorithms to so called memetic algorithms. The hope is that such algorithms inherent the good search properties of the strategies that they combine and are thus more universally applicable than strategies that focus solely on a search trajectory or on the combination through crossover.

10.4.1 Memetic Algorithms

Memetic algorithms combine the advantages of genetic algorithms with local search behavior. They are called "memetic" because the solutions are changed during the generation as opposed to only in the time between generations. The GLS algorithm mentioned in Section 10.1.2 is such a memetic algorithm. However, naturally one needs to take care that in the design of such an algorithm the population does not converge too quickly, as it draws its diversifying power from the genetic diversity in the population. The problem known as *genetic drift* could be even worse in this case when the local search decends into highly similar local optima. Here we compare the performance of GLS with that of a simple genetic algorithm that is adapted with local search.

In the results were the genetic algorithm was combined with a local search we can see that the search could indeed be improved over just the genetic algorithm or just the local search. The best-known solution was found only slightly more often than in the results of ILS (77.2% vs 76.6%), but the average quality could be significantly improved with an effort in between the first-improvement and best-improvement ILS. The algorithm did however not improve in all instances, but the improvement can mainly be attributed to instances where the ILS gave worse results such as `lip-a` and `tai-a`. It is interesting to see that GA+LS for example could

Table 10.8 Performance of GLS and GA+LS compared

	GLS			GA+LS		
Instance	Quality	Effort	Optimum	Quality	Effort	Optimum
bur	0.000%	114,751	100.0%	0.050%	343,540	87.5%
chr	0.009%	627,206	98.6%	0.269%	352,540	62.9%
els	0.000%	57,803	100.0%	0.649%	96,586	55.0%
esc	0.004%	91,802	99.4%	0.002%	80,480	99.7%
had	0.000%	19,718	100.0%	0.000%	16,114	100.0%
kra	0.000%	797,324	100.0%	0.132%	428,414	85.0%
lipa-a	0.000%	1,196,204	100.0%	0.300%	334,641	98.8%
lipa-b	0.000%	64,669	100.0%	0.000%	48,404	100.0%
nug	0.000%	166,994	100.0%	0.081%	130,326	87.7%
rou	0.000%	549,678	100.0%	0.259%	103,860	65.0%
scr	0.000%	56,966	100.0%	0.010%	59,681	98.3%
sko	0.081%	9,733,115	65.0%	0.367%	1,328,217	35.0%
ste	0.012%	2,905,395	91.7%	0.110%	743,858	55.0%
tai-a	1.720%	6,431,136	52.2%	3.077%	914,393	22.8%
tai-b	0.000%	403,488	100.0%	0.054%	403,152	86.3%
tho	0.042%	8,894,650	55.0%	0.268%	742,103	22.5%
wil	0.146%	17,108,610	40.0%	0.391%	1,778,433	10.0%
	0.158%	**1,444,752**	**93.1%**	**0.378%**	**334,290**	**77.2%**

not succeed in the els19 instance which the ILS perfectly solved. The hybrid inherited the problems of the genetic algorithm in this instance. GLS on the other hand shows good performance throughout the instances often solving them to optimality. Among all tested algorithms it did find the optimal or best-known solution most often.

10.5 Summary

Two algorithms achieved very good results overall: Robust taboo search (RTS) and genetic local search (GLS). However, the summarized results of the best performing metaheuristic for each problem instance family as shown in Table 10.9 suggest that for certain problem instances there are different algorithms that emerge as best. Algorithms in the table were deemed to perform better than another when they find on average better quality solutions as the maximum allowable effort in each algorithm was comparable. If two or more algorithms find equally good solutions, then the better performing algorithm is the one that is requiring less actual effort. Variable neighborhood search is not to be underestimated in situations where also the repeated or iterated local search is delivering good results. Also an implementation of tabu search that is oriented on the template described in [12] outperforms the robust variant on several instances. Usually both perform equally well, but the standard implementation uses less iterations. This shows the trade-off that robust taboo search makes in that it is on average performing much better. Finally, genetic local search showed the best overall results, but also emerged as the best algorithm only in some cases.

Table 10.9 Summary table that lists the best performing results

Instance	Best performing	Instance	Best performing
bur	VNS	rou	Standard TS
chr	Genetic LS	scr	Standard TS
els	VNS	sko	Genetic LS
esc	Genetic LS	ste	Genetic LS
had	VNS	tai-a	Robust TS
kra	Genetic LS	tai-b	Genetic LS
lipa-a	Standard TS	tho	Robust TS
lipa-b	Standard TS	wil	Robust TS
nug	Robust TS		

10.6   Conclusions

As we have shown in this work the quadratic assignment problem is an interesting problem despite its age. Some standard metaheuristics from the first days do not perform as well and several modifications need to be made. Metaheuristics of later generations such as variable neighborhood search do perform very well out of the box, but the clever combination of population-based and trajectory-based approaches can lead to very good results overall. It is interesting that the diversity in the characteristics of these instances is reflected in the heterogeneity of the applied algorithms and while there exist overall well-performing algorithms they are often not the best algorithm for every instance. The results also suggest that, despite the overall strength of RTS and GLS there is no single best algorithm for all the instances. To design better algorithms is one possible approach to continue making progress in solving the QAP, but more importantly, research needs to be done to decide which algorithm to choose for a concrete and previously unobserved instance. As more experiments are performed in such studies the knowledge on the performance on individual algorithms is increasing. The topic of fitness landscape analysis seems to be most promising to extract problem instance characteristics that could be used to indicate which algorithm to choose and constitutes an interesting base for further research.

Acknowledgements. The work described in this chapter was done within the Josef Ressel-Centre HEUREKA! for Heuristic Optimization sponsored by the Austrian Research Promotion Agency (FFG).

References

1. Affenzeller, M., Wagner, S.: Offspring selection: A new self-adaptive selection scheme for genetic algorithms. In: Ribeiro, B., Albrecht, R.F., Dobnikar, A., Pearson, D.W., Steele, N.C. (eds.) Adaptive and Natural Computing Algorithms. Springer Computer Series, pp. 218–221. Springer (2005)
2. Affenzeller, M., Winkler, S., Wagner, S., Beham, A.: Genetic Algorithms and Genetic Programming - Modern Concepts and Practical Applications. Numerical Insights. CRC Press (2009)
3. Angel, E., Zissimopoulos, V.: On the landscape ruggedness of the quadratic assignment problem. Theoretical Computer Science 263(1-2), 159–172 (2001)
4. Burkard, R.E., Karisch, S.E., Rendl, F.: QAPLIB - A quadratic assignment problem library. Journal of Global Optimization 10(4), 391–403 (1997)
5. Chicano, F., Luque, G., Alba, E.: Autocorrelation measures for the quadratic assignment problem. Applied Mathematics Letters 25, 698–705 (2012)
6. Czech, Z.J.: Statistical measures of a fitness landscape for the vehicle routing problem. In: Proceedings of the 22nd IEEE International Parallel and Distributed Processing Symposium, IPDPS 2008 (2008)
7. de Carvalho Jr., S.A., Rahmann, S.: Microarray layout as quadratic assignment problem. In: Proceedings of the German Conference on Bioinformatics (GCB). Lecture Notes in Informatics, vol. P-83 (2006)
8. Drezner, Z.: Extensive experiments with hybrid genetic algorithms for the solution of the quadratic assignment problem. Computers & Operations Research 35, 717–736 (2008)
9. Drezner, Z., Hahn, P.M., Taillard, E.D.: Recent advances for the quadratic assignment problem with special emphasis on instances that are difficult to solve for meta-heuristic methods. Annals of Operations Research 139, 65–94 (2005)
10. Elshafei, A.N.: Hospital layout as a quadratic assignment problem. Operational Research Quarterly 28(1), 167–179 (1977)
11. Fogel, D.: An evolutionary approach to the traveling salesman problem. Biological Cybernetics 60, 139–144 (1988)
12. Glover, F.: Tabu search – part I. ORSA Journal on Computing 1(3), 190–206 (1989)
13. Hahn, P.M., Krarup, J.: A hospital facility layout problem finally solved. Journal of Intelligent Manufacturing 12, 487–496 (2001)
14. James, T., Rego, C., Glover, F.: Multistart tabu search and diversification strategies for the quadratic assignment problem. IEEE Transactions on Systems, Man and Cybernetics, Part A: Systems and Humans 39(3), 579–596 (2009)
15. Koopmans, T.C., Beckmann, M.: Assignment problems and the location of economic activities. Econometrica, Journal of the Econometric Society 25(1), 53–76 (1957)
16. Merz, P., Freisleben, B.: A genetic local search approach to the quadratic assignment problem. In: Proceedings of the Seventh International Conference on Genetic Algorithms, pp. 465–472. Citeseer (1997)
17. Merz, P., Freisleben, B.: Fitness landscape analysis and memetic algorithms for the quadratic assignment problem. IEEE Transactions on Evolutionary Computation 4(4), 337–352 (2000)
18. Pitzer, E., Affenzeller, M., Beham, A.: A closer look down the basins of attraction. In: UK Conference on Computational Intelligence (2010) (in press)
19. Stutzle, T.: Iterated local search for the quadratic assignment problem. European Journal of Operational Research 174, 1519–1539 (2006)
20. Taillard, E.D.: Robust taboo search for the quadratic assignment problem. Parallel Computing 17, 443–455 (1991)

Chapter 11
TV-Anytime Cloud Computing Concept in Modern Digital Television

Artur Bąk and Marek Kulbacki

Abstract. With the increase in the demand for digital television (DTV) in various standards, high quality content (HDTV), mobile TV, video-on-demand services and interactive TV, a new problem of much more complexity in various TV technologies appeared that makes cross platform solutions very difficult to achieve. One of the main factors increasing complexity is wide use of the Internet for delivering such TV services beside the traditional broadcast of TV content. To make the connection of these technologies feasible and in this way to provide the TV users with easy and convenient access to TV content from as many sources as possible some kind of unification and standardization needed to be introduced. This paper provides an overview of global standardization approach, namely the TV-Anytime (TVA) standard [6]. The aim of the TVA standard is to enable a flexible use of TV across a wide range of networks and connected devices. One of the main assumptions of TVA is that user should be fully agnostic of technology and source of the requested content in the same way as in cloud computing domain. We describe the motivation of its use, its principles as well as the base features defined by this standard to satisfy the most recent TV domain trends. We provide also the practical example of the TVA system usage for better understanding the theoretical principles.

11.1 Introduction: Trends for TV-Anytime and TV-Anywhere

Nowadays the television (the commercial as well as non-commercial one) available for regular recipient differs significantly from that existing a decade or two in the past. The change from the analogous television to the digital one enabled a lot of new pos-

Artur Bąk
e-mail: abak@ftoday.net

Marek Kulbacki
Polish-Japanese Institute of Information Technology,
Koszykowa 86, 02-008 Warszawa, Poland
e-mail: kulbacki@pjwstk.edu.pl

© Springer International Publishing Switzerland 2015   191
R. Klempous and J. Nikodem (eds.), *Innovative Technologies in Management and Science*,
Topics in Intelligent Engineering and Informatics 10, DOI: 10.1007/978-3-319-12652-4_11

sibilities as better quality or transmitting more channels in the similar infrastructure (e.g. on the same frequency of given satellite transponder). But probably the most important new feature of the innovative digital television is the possibility to transmit the additional data and services together with audio/video content that allows deliver it to the recipient in much more flexible and sophisticated way than in the past [1]. The crucial application of that is delivering so called "non-linear" content to the user.

To explain the difference between linear and non-linear content let us imagine the standard way of watching television. The television broadcaster (the company delivering the content to user) transmits the event of particular channel on the specific transponder on the specific frequency at specific time. At that specific time the user can tune his TV receiver (TV set or set-top-box) to that specific frequency and watch the event. Although the access to the schedule for events is now public and easy (as a basic feature of each modern TV set or set-top-box) the major inconvenience is that user must be in front of his device at that specific transmission time. If this is not possible at that time the user must miss the event and wait till broadcaster will transmit it again. As an opposite, the non-linear content is delivered to the user at the users explicit request at the users best convenient moment. Such way of delivering content to the user is also commonly called video-on-demand (VOD). There are various ways of delivering VOD. One of the options is pushing content to the users receiver by storing it on the internal storage (e.g. HDD) without user awareness. After content is stored on receiver, the user is offered to select content to watch. Such scenario is called push-VOD. Another way is the direct streaming of the content at the users request without storing it on internal storage (in other word the content is being consumed at the moment of the content transmission). The difference between linear content and streaming is that for the latter one the content is dedicated for particular user at the time selected by that user. The "in-between" solution for linear and non-linear content is the Personal Video Recording (PVR) which relies on manual recording of linear content on the internal storage of users receiver exactly at the time while linear content is being transmitted. The operation of recording the content is triggered by user and once completed the recoded content is available locally on receiver so that the user could watch the content at any time he wants. The special version of PVR is its network form which is called Network PVR, recently developed also in cloud model [2, 3]. In this case the user initiates the recording of content as usual PVR but the content is stored on the remote server in the network. At any time the user wants to watch the content, the streaming is initiated and user consumes the content directly from the network. There are many different varieties of VOD and PVR solutions but here there were mentioned only basic ones, just for general feeling of the new way of content delivery and consumption time. The common advantage of all these new ways of non-linear content delivery over linear content is the possibility to pause, rewind and forward the playback of content with different rate (e.g. from x2 to x64 times as normal playback). These actions on playback are commonly called "trick-plays".

Another change introduced by recent digital technology is strong role of the Internet in the content delivery. One or two decades before there were only three basic carriers of transmissions: terrestrial, satellite and cable broadcast . At the end of last century there was usually only on-way communication broadcast of content from broadcaster to users receiver . More less ten years ago the IP connection started to be commonly available on the set-top-boxes (and not so often on TV sets). Initially this possibility (called "return channel") was only used for simple functions as register-ing the customers of broadcaster or simple feedback from customers. But during last ten years, the IP connection in digital TV industry became to play the crucial role in content delivery. It is not any more the simple "return channel" of receivers but can completely replace the standard carriers of transmission (beside the terrestrial, satellite or cable broadcast). The transmission of content (or any accompanying data) by IP is called "broadband" to differentiate it from traditional "broadcast". The one strong advantage of broadband (comparing to the broadcast) is that it is bi-directional by default so the TV operator has usually the similar capabilities of transmission of data (or content) to the end user as the user to the opposite direction.

The third change in TV industry is possibility to watch the delivered content on more kinds of devices than before. Nowadays apart from traditional TV sets or set-top-boxes, the content can be handled on PC computers, tablets and even on mobile phones. Additionally the progress in the Internet providing technology allows the content to be easily caught in almost every causal place the user is currently stay-ing. This is strong advantage comparing to the traditional watching the television at home in the past.

Described above those three big changes in television industry enabled the flex-ible and convenient mean to consume the content but on the other hand they intro-duced a lot of complexity to the whole TV world. Especially that the current trends in TV industry (driven mostly by users demands) require to consume content any-time and anywhere. These two last concepts (any-time and anywhere) became the official name of technology and means respectively watching content any time the user wants and on any device the user has currently access to (TV set, tablet, mobile or any other device capable of viewing the content).

To make the any-time and anywhere concepts feasible in practice the different technologies need to find the "common language" to communicate about the con-tent in similar way [5, 4]. In other words the common standardization needs to be proposed. In the rest of this article we focus mostly on solutions proposed for any-time concept having in mind that this must be valid also for any device to satisfy the any-where concept.

The main assumption of standardization described in this chapter is that user should be agnostic of the location of the content and the time when that content is available (there can be no even plan in the broadcaster schedule transmission of this content in the future). The user just wants to see the content (by simple and human-readable indication) on the casual device he owns at the moment. Once the system has an access to such content it will offer it to the user for watching. Such approach is similar to the cloud concept in standard computing and data management domains.

The following sections provide the general overview of main concept introduced by TV-Anytime standard. It provides the way of treating content in unified way in different environments. First of all the basic principles of TV-Anytime standard are described with pointing new functions that standard requires from its implementations to follow the current market needs. Additionally the practical example of TV-Anytime use-case is presented to allow the reader for better understanding how such solution can work in practice.

11.2   Principles of TV-Anytime Standard

In general the TV Anytime standard is set of complementing specifications defining the search, selection, acquisition and consumption of content delivered to the user by broadcast or broadband (online) services. The main purpose of this is to propose the context and architecture that should be implemented by systems that are intended to work properly in TV Anytime environment.

The following sub-sections describe two basic models of TV-Anytime system: the simple broadcast model as well as more complex fully interactive one. Next, the basic TV-Anytime functions such content referencing and metadata are described in details. At the end the general usage template of TV-Anytime system is defined.

11.2.1   Simple Broadcast Model

In the simplest TV-Anytime system offering just the broadcast of content, there can be identified three major elements:

- service provider delivers TV-Anytime service;
- transport provider carries the service
- users equipment stores the content and plays it back for the user;

These major elements must commonly realize the set of functions that can be defined as follows:

- search and navigation;
- location resolution;
- user interaction
- content presentation;
- local storage management;

The simple broadcast model is depicted on the Fig. 11.1. Each component on that diagram represents the particular functions of the TV-Anytime system and can be implemented in different ways in practice. The connections between the components represent the information flows between the functions. This simple broadcast model represents typical PDR receiver - the device (e.g. at users home) for recording and view the content. Such device receives the content in traditional way (by broadcast)

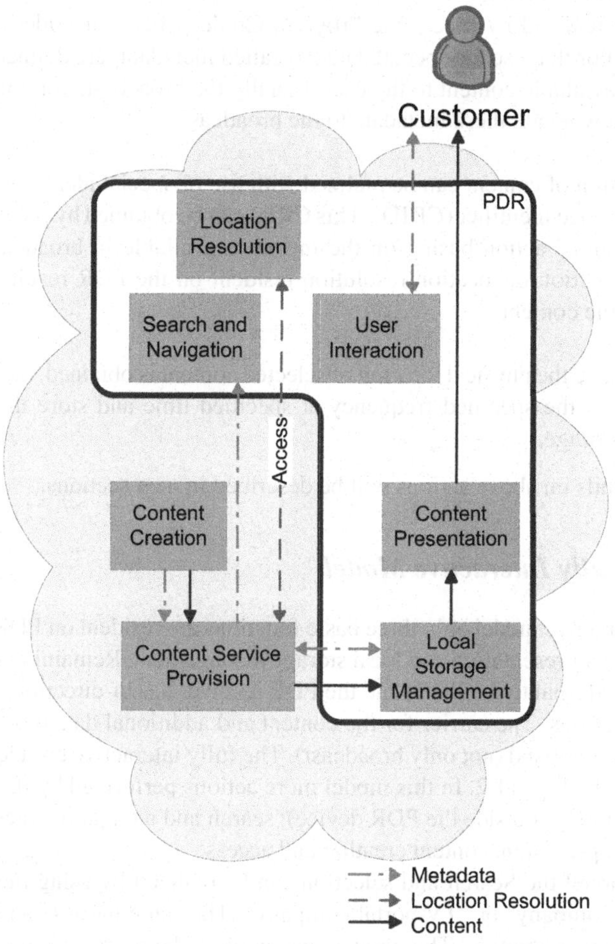

Fig. 11.1 Simple broadcast model of TV-Anytime system

and stores it on the local storage. Stored content can be later played back at the users request.

The only functions realized outside the PDR are: content creation, content service provision and access. The remaining functions are embedded (resident) on the PDR device.

As consequence, with use of above elements of the system and functions they offer, the user is able to perform following actions (before the content will be ready to be viewed):

Search → Select → Locate → Acquire → *content is ready for watching*

Below there is description how TV-Anytime standard propose to realize each of above actions.

- **Search and selection**

  It can be realized by Electronic Program Guide (EPG) embedded on the PDR receiver. For that use the special data (so called metadata) are defined that advertises the available content to the user. Usually the broadcaster or content creator is the one who add these metadata to the broadcast.

- **Locate**

  The location of content can be realized with use of special identifier called Content Reference Identifier (CRID). This CRID can be obtained by previous "Search and selection" action basing on the metadata available in broadcast [11]. The special function of location resolution resident on the PDR results in physical location the content.

- **Acquire**

  Finally, once the physical location of selected content is obtained, the PDR device can tune to the specified frequency at specified time and store the content on internal storage.

More details on above actions will be described in next sections.

11.2.2 Fully Interactive Model

In fully interactive model only three basic functions are resident on PDR: user interaction, content presentation and local storage management. Remaining function can be remote (external to the PDR) and the PDR receiver has bi-directional connection to these functions. The carrier for the content and additional data transmission can be also the broadband (not only broadcast). The fully interactive broadcast model is depicted on the Fig. 11.2. In this model more actions performed by the user can be realized remotely (outside the PDR device): search and navigation, location resolution, content provision, content creation and access.

In this model the Search and selection can be realized by using the specialized "Web-EPG company" or "TV-portal company". The same parties can be responsible by location resolution. This process can assume the users awareness of remote functionality or can be fully transparent for the user. Such solution can be treated as cloud computing concept adopted onto the digital television domain.

11.2.3 Content Referencing and Location Resolution

The very important capability of TV-Anytime system is that user can select the content regardless of its location or time of availability. Just to illustrate this concept better let us imagine that the user sees some movie advertisement and PDR receiver allows user to record the content for later viewing. At the moment of recoding request the exact time and location of transmission for such movie can be unknown to PDR receiver. Moreover, even a broadcaster may have not it yet in the schedule. Anyway, the TV-Anytime system should ensure that user can request such movie at that time and do not miss it once this content became available on any broadcast channel or even on Internet file server (if PDR offers such feature).

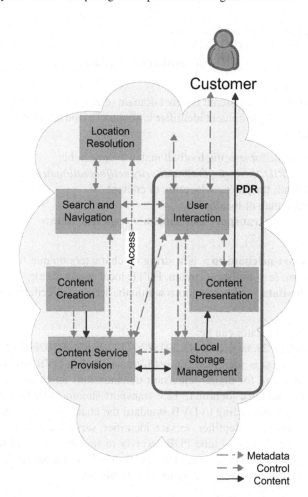

Fig. 11.2 Fully interactive model of TV-Anytime system

The user can request the recording also the whole series of episodes. In this case the PDR should obtain the time and location for each particular episode of the series and acquire them once available. For that reason the content referencing tool must be flexible enough to point single content and collection of contents.

The main concept of content referencing defined by TV-Anytime is the separation of reference to content from specific information which is necessary to physically acquire the content. It allows to indicate content before the final location and time of content transmission is available (as only unique reference to content is necessary at that time). Additionally such approach allows to map the single content reference to many physical locations and improves the efficiency of the system (alternative location can be used for the same content obtained with the same content reference).

To provide the separation of two elements of content referencing mentioned above there were distinguished:

- **CRID** used as reference to the content
  The general syntax of CRID is :
  $$\textbf{CRID://<authority>/<data>}$$
  where:

  - <**authority**> is registered Internet domain (e.g. *www.football-authority.net*);
  - <**data**> is unique content identifier in scope of given authority (e.g. */footbal-l/match10*).

  The full example for specific football match event can be:
  $$CRID://www.football-authority.net/football/match10.$$

- **Locator** defines the physical location of content
  The general syntax of locator is :
  $$\textbf{<transport-mechanism>:<specific-data>}$$
  where:

  - <**transport-mechanism**> is a string of characters unique for transmission mechanism (e.g. transport stream, HTTP, local storage file);
  - <**specific-data**> is unambiguous within the scope of specific transport mechanism.

  The example of locator can be:
  $$dvb://123.5ac.3be;3e45\ 20131207T120000Z–PT02H10M$$
  where the content is identified by *3e45*, transmitted in DVB network (DVB is European standard for digital television transmission) on channel identified by *123.5ac.3be* (network identifier: 123, transport stream identifier: *5ac*, service identifier:*3be*). According to DVB standard the channel is referenced by three parameters: network identifier, service identifier, service identifier. These three parameters are enough to tune PDR receiver to specified channel in DVB network. The planned date of content transmission is 7th of December 2013 at 12:00 am.The duration of content is 2 hours and 10 minutes.

The content identifier CRID is created at the time of content creation and should stay constant and unique for particular content for ever. The location resolution process converts the CRID into the locator or other CRID-s (in case of series) as illustrated on Fig. 11.3.

The CRID itself is not enough for PDR receiver to resolve the location (and find the locator as a result). Therefore additional data must be delivered to the PDR [9, 10]. These data are provided as two well defined XML data structures called "tables" (in digital television the data structures sent together with the content in the same carrier are usually called "tables"):

- **RAR Table** (Resolving Authority Record) It maps the authority that issued the CRID to resolution service provider.
- **Content Referrencing Table** It is the actual resolution table which maps the CRID to final locator or another CRID-s (for series).

The RAR Table contains one or more resolution service providers for each authority that submits CRID. The example of one entry RAR Table is shown on

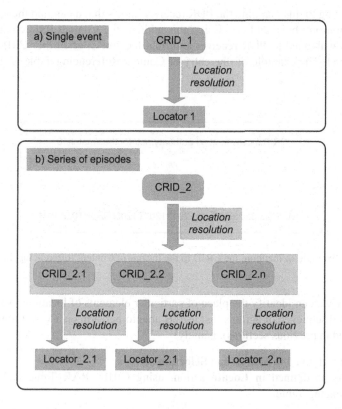

Fig. 11.3 Location resolution process for (a) single event (e.g. one movie) and (b) for series of episode (that can be transmitted at different time and location)

Fig. 11.4. For previously mentioned authority *www.football-authority.net* the resolution service provider is footballresloc.net and the Internet address to that provider can be *http://footballresloc.net/resloc*.

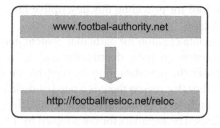

Fig. 11.4 Example of mapping the authority to the resolution service provider in RAR Table

According to this mapping the PDR receiver knows that it can find the resolution information on above address. Using this address the Content Referencing Table can be downloaded to PDR receiver and final locator(-s) or further CRID(-s) can be obtained. The example of one entry of Content Referencing Table is shown on Fig. 11.5.

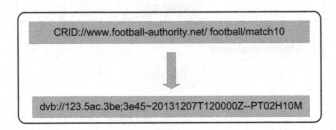

CRID://www.football-authority.net/ football/match10

dvb://123.5ac.3be;3e45~20131207T120000Z--PT02H10M

Fig. 11.5 Example of mapping the CRID to the locator in Content Referencing Table

Finally we can identify the place of particular elements of location resolution in the general sequence of actions that user is able to perform on TV-Anytime system (described in previous sections), namely:

- CRID is defined in **Search** and **Selection** actions;
- Locator is defined in **Locate** action, using CRID, RAR Table and Content Referencing Table.

11.2.4 Metadata

Before selecting the particular content the user needs to have some descriptive information on that content to make decision. The user is usually interested in title, gender (e.g. drama or comedy), actors or plot description. Only basing on this the user can select content so that PDR could obtain specific CRID. On the other hand the content creators and broadcasters use such information to attract the user for their content. The tool for providing this descriptive information is concept of content-related metadata.

The process of metadata creation for particular content may involve many organizations during the content creation, distribution and delivery to the user phase. After delivery to the user the metadata must be readable on different devices and software technologies. For that reason there is a need of defining common metadata framework to ensure high level of interoperability on different usage phases and on different technologies. The metadata format adopted by TV-Anytime standard is XML that ensures extensibility, being agnostic of application technology and wide usage around different domains and technologies [8].

There are following basic kinds of metadata:

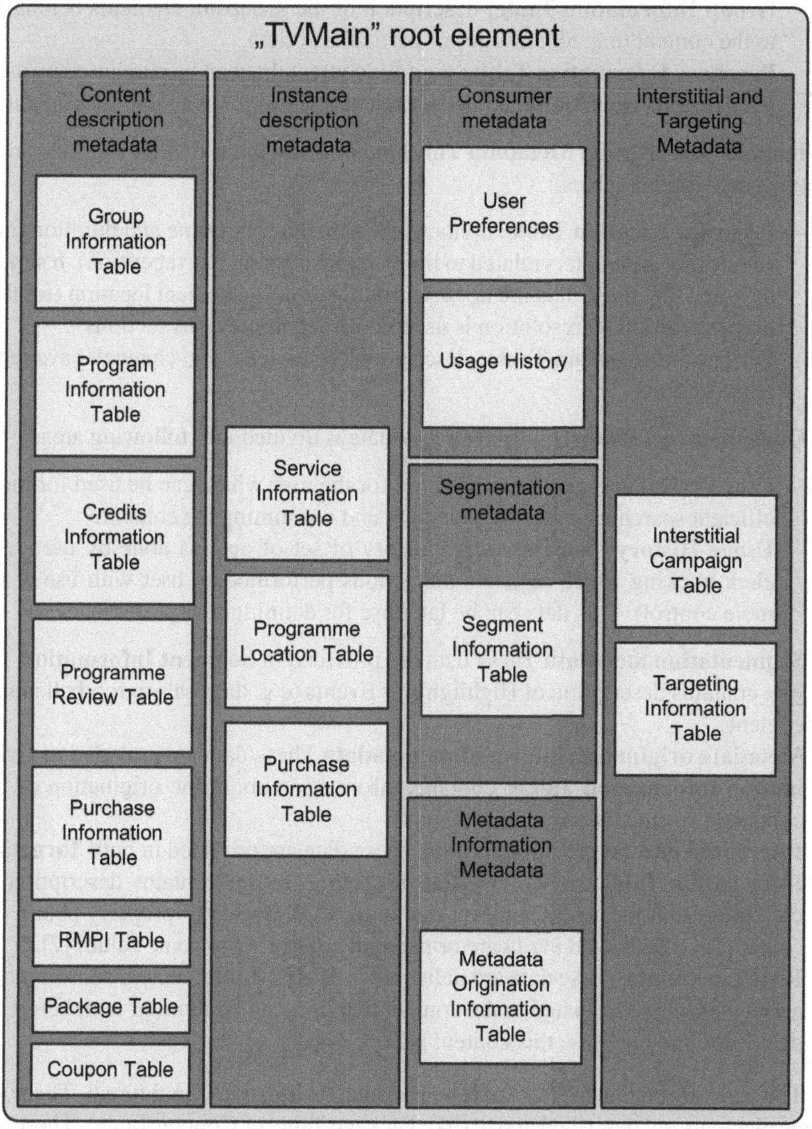

Fig. 11.6 TV-Anytime documents with "TVA Main" as a root element

- **Content description metadata** This kind of metadata is divided into four areas (each area is delivered in specific structure called "Table"):

  - **Program Information Table:** description of elements belonging to the content it can include a title of content, gender, keywords etc. These data can be used for search content by user.

- **Group Information Table:** description of the groups of elements belonging to the content (e.g. all episodes of particular series).
- **Purchase Information Table:** used for content that can be purchased by user.
- **Program Review Table:** contains critical reviews for available for content.

- **Instance Description Metadata** This kind of metadata is divided into two areas (also delivered in tables):

  - **Program Location Table:** contains the schedule start time and duration (and some other parameters related to time of transmission e.g. repetition). It should be noted that these data are not used for determining content location (for that purpose the CRID resolution is used described in previous sections).
  - **Service Information Table:** description of services (e.g. channels) available in the system.

- **Consumer metadata** This kind of metadata is divided into following areas:

  - **User Preferences:** contains a profile for the user which can be used for more efficient searching, filtering, selecting and consuming the content.
  - **Usage History:** contains usage history of set of actions done by user (e.g. click tracking which registers the actions performed by user with use of remote control). This data can be later use for defining user preferences.

- **Segmentation metadata** These data are provided in **Segment Information Table:** contains description of Highlights or Events (e.g. the goals in football match content).
- **Metadata origination information metadata** These data are provided in **Origination Information Table:** contains information about the origination of the content (e.g. cinema originated content).
- **Interstitial and targeting metadata** These data are provided in both **Targeting Information Table** and **Interstitial Targeting Table:** contains description of conditions that must be satisfy on the target PDR device to properly obtain the content (like dedicated hardware or parental control set to some value) [7].
- **RMPI metadata** These data are delivered in **RMPI Table:** contains information about the rights associated to the content that user can be familiar with before he decided to the purchase this content [12].

The general schema of TV-Anytime metadata is illustrated in figure 6. There are some elements not described above (like Package Table or Coupon Table). These additional elements will be mentioned later in this chapter in description of additional features supported by TV-Anytime metadata.

Generally the main goal of metadata is to provide the user with tool for content selection as well as to provide the service provider (content creator or broadcaster) the tool for attracting the user. Apart from that the TV-Anytime standard defines some additional features that metadata must supported to keep the most recent trends in digital television like describing user profiles, search preferences or event facilitating the filtering and acquisition of content in behalf of user by automatic agents. The list of features that metadata must support to make above tasks feasible is presented below:

- New content types;
- Packaging;
- Targeting;
- Interstitial content;
- Sharing;
- Remote programming;
- Coupons.

11.2.5   Additional Features Supported by Metadata

In this section the additional features supported by TV-Anytime metadata are described.

New Content Types

The metadata should support other content than standard audio-video content. The examples of such content are: graphics files, music files, web-pages, video-games or even interactive applications.

Packaging

Metadata should enable the combination of different content types as audio-video, games, applications or images. All components of the package are intended to be consumed together. The example of packaging can be the additional soccer game combined together with the football match content plus some additional still images of particular players attending the game. Package description metadata may provide also the options (information which package components should be consumed if not all) and time synchronization between them. The example can be the multi-camera feature for sport events.

Targeting

Metadata enable automatic matching the relevant content to profiled customers. There are two types of targeting:

- **Push targeting:** broadcaster delivers content that can be consumed depending on the user profile;
- **Pull targeting:** intelligent agent selects the content depending on user profile and stores it on users PDR receiver.

Generally targeting uses the user preferences and usage history that can be stored locally on PDR receiver or on remote server.

Interstitial Content

Metadata enable interruption mechanism to replace the currently being played back content with another one. The example of such interrupting content is advertisement spot transmitted during the playback of movie.

Sharing

Metadata enable users to notify other users about interesting content. Additionally it provides the mean to manage the configuration of alternative PDR device according to users profile. Also the sending content to other devices or to other users is considered by such kind of metadata.

Remote Programming

Metadata enable users to program the recording of content from other device that actually records the content. The example of this feature is requesting the recording on home PDR receiver from mobile phone being physically outside home (e.g. in the office).

Coupons

Coupons metadata provides the concept of electronic value that can complement or replace the money during content purchase operation. The examples of coupons are discounts, "two products instead of one" promotion or "buy two get three" promotion.

11.2.6 Template of TV-Anytime Usage

Having all tools for location resolution defined we can now define the general template of TV-Anytime usage realizing the sequence of actions defined above (see Fig. 11.7).

Publish

The service provider (content creator or broadcaster) generates the CRID and metadata for content that will be available for the user.

Search

User searches the content using some kind of EPG tool (Electronic Program Guide) embedded in PDR receiver or remote Web EPG tool. Each kind of EPG tool (embedded or remote) provides the descriptive information about available content extracted from content-related metadata.

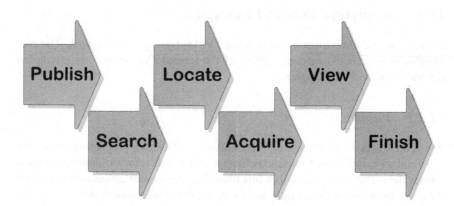

Fig. 11.7 Sequence of consecutive actions in TV-Anytime system usage

Select

User selects the content and content CRID is identified.

Locate

PDR receiver performs location resolution:

- It uses RAR Table to identify the resolution service provider (using authority from CRID);
- It obtains the locator using Content Referencing Table provided by resolution service provider.

Acquire

The content is downloaded according to the locator and stored on PDRs local storage once it is available.

View

Content is ready for users consumption.

Finish

After consumption of content by user the usage history can be stored on the local storage of PDR receiver or can be sent to the server. Using this data the user preferences can be defined and used later for enhancement of the user experience during his searching the content next time.

11.3   TV-Anytime Practical Example

In this chapter there is an example how the TV-Anytime system can work. The consecutive steps (actions) are compatible with the general TV-Anytime usage template defined in the previous chapter.

11.3.1   Publish

The content creator or broadcaster generates new CRID after production of the content football match from Champions League 2013. The same or another company creates the metadata that describes this football event as well as location resolution data with information on the time and location of content transmission.

The constructed metadata describing the content (Program Information Table) may look like XML snippet below:

```
<ProgramDescription>
 <ProgramInformationTable>
   <ProgramInformation programId="crid://sport-authority.com
       /football/match10">
     <BasicDescription>
        <Title type="main">Dortmund Vs Bayern</Title>
        <Synopsis length="short">Semi-final of Champions
            League belongs to Germany
        </Synopsis>
     </BasicDescription>
   </ProgramInformation>
 </ProgramInformationTable>
 <ProgramLocationTable>
   <BroadcastEvent serviceIDRef="hbc10022311">
      <Program crid="crid://ˍsport-authority.comˍ/football/
          match10" />
      <ProgramURL>dvb://1.4ee2.3f4/</ProgramURL>
      <PublishedStartTime>2013-05-25T18:00:00.00+01:00Z</
          PublishedStartTime>
      <PublishedDuration>PT6H</PublishedDuration>
   </BroadcastEvent>
 </ProgramLocationTable>
</ProgramDescription>
```

11.3.2   Search

User is searching the content for watching using EPG embedded on the PDR receiver. The PDR uses Program Information Table and Program Location Table to render the EPG. Using the EPG the user can navigate and search the interesting

```
<UserDescription >}
 <UserPreferences >
   <mpeg7: UserIdentifier protected="true">
     <mpeg7:Name xml: lang="en">John Simpson </mpeg7:Name>
   </mpeg7: UserIdentifier >
   <mpeg7: FilteringAndSearchPreferences >
     <mpeg7: ClassificationPreferences preferenceValue="12">
     <mpeg7: Language >en </mpeg7: Language >
     <mpeg7: Genre href="urn: tva: metadata: cs:
     ,FormatCS:2007:3.2.3.12"/>
     <mpeg7: Subject >Football </mpeg7: Subject >
     </mpeg7: ClassificationPreferences >
   </mpeg7: FilteringAndSearchPreferences >
 </ UserPreferences >
</ UserDescription >
```

contents. One of the ways of searching can be that user types 'champions league' as a keyword in EPG. As a result the EPG can examine the title and synopsis fields of all available contents and offers the user the one that matches the requested keyword: Dortmund Vs Bayern: Semi-final of Champions League belongs to Germany

If the user used to watch many football games before, the EPG can suggest the selection of above event to him without typing the keyword by user. In this case the following metadata describing user preferences can be delivered to PDR:

11.3.3   Select

Once user made a selection the system can start to realize his request. At this time the usage history can be updated and stored on PDR local storage or send it to remote server managing user profiles. From now on the system will be tracking the availability of content and acquire it once available.

11.3.4   Locate

After selection of the content the location must be resolved. The CRID is extracted from Program Information Table and use for defining the locator using Content Referencing Table.

The Content Referencing Table for above football match can look like a metadata snippet below:

```
<ContentReferencingTable>
<!-- CRID resolution to locators -->
 <Result CRID="crid://sport.com/football/match10"
 status="resolved" complete="true" acquire="all">
  <LocationsResult>
   <Locator>dvb://1.4ee2.3f4;4f5~2013-05-25T18
    :00:00.00+01:00/PT01H30M
   </Locator>
  </LocationsResult>
 </Result>
</ContentReferencingTable>
```

For requested football match the only one DVB locator is defined. Also the exact time and duration is known at the moment of content resolution.

There can be also different scenario when exact time of content transmission is not known yet and Content Referencing Table snippet can look like this:

```
<ContentReferencingTable>
<!-- CRID resolution to locators -->
 <Result CRID="crid://sport.com/football/match10"
 status="cannot_yet_resolve" complete="true"
 acquire="all" reresolveDate="2013-05-20T12:00:00.00+01:00"
    ></Result>
</ContentReferencingTable>
```

11.3.5 Acquire

Once the start time of content (if the location is already resolved) will be reached the PDR receiver will tune to the specified channel and start recording of content. When recording is complete the content is announced to be available for viewing by PDR receiver. In case when location cannot be resolved the acquiring process must be postponed till the location resolution is possible.

11.3.6 View

User can view the recorded content on PDR receiver once the recording is complete. Depending on the system available on PDR the user can send critical review of the content back to the broadcaster, recommend the content to his friends or send it to them. It is possible to also send the content to other devices (like tablets or smart phones) if PDR system offers such functionality.

11.3.7   Finish

After consumption of content the usage history can be updated for that user. For example the recording operation itself can be add to the usage history and following metadata can be generated and store in local storage or send back to the broadcaster:

```
<UserDescription>
 <UsageHistory id="usage-history -001" allowCollection="true"
    >
<mpeg7:UserIdentifier protected="true">
 <mpeg7:Name xml:lang="en">John Simpson</mpeg7:Name>
</mpeg7:UserIdentifier>
<mpeg7:UserActionHistory id="useraction-history -001"
    protected="false">
 <mpeg7:ObservationPeriod>
  <mpeg7:TimePoint>2013-05-25T6:00 -23:00</mpeg7:TimePoint>
 </mpeg7:ObservationPeriod>
 <mpeg7:UserActionList id="ua-list -001" numOfInstances="1"
  totalDuration="PT1H30M">
  <mpeg7:ActionType href="urn:tva:metadata:cs:ActionTypeCS
    :2004:1.3">
   <mpeg7:Name>Record</mpeg7:Name>
  </mpeg7:ActionType>
  <mpeg7:UserAction>
   <mpeg7:ActionTime>
    <mpeg7:MediaTime>
    <mpeg7:MediaTimePoint>2013-05-25T18:00:00</mpeg7:
       MediaTimePoint>
     <mpeg7:MediaDuration>PT1H30M</mpeg7:MediaDuration>
    </mpeg7:MediaTime>
   </mpeg7:ActionTime>
   <mpeg7:ProgramIdentifier organization="TVAF"
   type="CRID">://sport.com/football/match10
   </mpeg7:ProgramIdentifier>
  </mpeg7:UserAction>
 </mpeg7:UserActionList>
</mpeg7:UserActionHistory>
 </UsageHistory>
</UserDescription>
```

11.4   Summary

The TV-Anytime standard seems to be good solution for introducing the unification method which can make the global distributed system for TV consumers that can select and acquire TV content from the global cloud system. The TVA standard is developed and supported by many significant organizations acting on DTV domain which is a huge advantage in adaptation it to TV market process. Although its proven right direction and support by key TV market players the full adaptation of the TVA standard globally is not easy task. There are too many of existing legacy solutions that are not prepared for adjusting all standard requirements. For that reason most TV service providers use still their own proprietary solutions for TV anytime and TV anywhere features and introducing one global standard seems to be long way process. In this document we describe only the principles of TV-Anytime standard and simple examples to get the reader familiar with general picture of TVA. To go into more detailed specification or to use the standard in practice the reader should study the references indicated in this document especially [6, 7, 8, 9, 10, 11, 12].

References

1. El-Hajjar, M., Hanzo, L.: A survey of digital television broadcast transmission techniques. IEEE Communications Surveys and Tutorials (2013)
2. Alsaffar, A.A., Huh, E.: A framework of N-screen services based on PVR and named data networking in cloud computing. In: Proceedings of the 7th International Conference on Ubiquitous Information Management and Communication (ICUIMC 2013), New York, USA (2013)
3. Roüast, M., Bainbridge, D.: Live television in a digital library. In: Proceedings of the 12th ACM/IEEE-CS Joint Conference on Digital Libraries (JCDL 2012), New York, USA (2012)
4. Solla, A.G., Bovino, R.: Tv-Anytime: Paving the Way for Personalized TV. Springer Publishing Company, Incorporated (2013)
5. Soares de Oliveira, F., Batista, C.E., de Souza Filho, G.L.: A^{3}TV: anytime, anywhere and by anyone TV. In: Proceedings of the 12th International Conference on Entertainment and Media in the Ubiquitous Era, MindTrek 2008 (2008)
6. ETSI TS 102 822-2 V1.4.1 (2007-11): Broadcast and On-line Services: Search, select, and rightful use of content on personal storage systems ("TV-Anytime"); Part 2: Phase 1 - System description
7. ETSI TS 102 822-3-4 V1.6.1 (2012-12): Broadcast and On-line Services: Search, select, and rightful use of content on personal storage systems ("TV-Anytime"); Part 3: Metadata; Sub-part 4: Phase 2 - Interstitial metadata
8. ETSI TS 102 822-3-1 V1.8.1 (2012-12): Broadcast and On-line Services: Search, select, and rightful use of content on personal storage systems ("TV-Anytime"); Part 3: Metadata; Sub-part 1: Phase 1 - Metadata schemas
9. ETSI TS 102 822-4 V1.7.1 (2012-12): Broadcast and On-line Services: Search, select, and rightful use of content on personal storage systems ("TV-Anytime"); Part 4: Phase 1 - Content referencing
10. ETSI TS 102 323 V1.5.1 (2012-01): Digital Video Broadcasting (DVB); Carriage and signalling of TV-Anytime information in DVB transport streams
11. Aoki, S., Ashley, A., Kameyama, W.: The TV-Anytime Content Reference Identifier, CRID (2005), ftp://ftp.rfc-editor.org/in-notes/rfc4078.txt
12. ETSI TS 102 822-5-2 V1.2.1 (2006-01):Broadcast and On-line Services: Search, select, and rightful use of content on personal storage systems ("TV-Anytime"); Part 5: Rights Management and Protection (RMP) Sub-part 2: RMPI binding

Subject Index

Printed in the United States
By Bookmasters